Application of Advanced Optimization Techniques for Healthcare Analytics

Application of Advanced Optimization Techniques for Healthcare Analytics is an excellent compilation of current and advanced optimization techniques which can readily be applied to solve different hospital management problems. The healthcare system is currently a topic of significant investigation to make life easier for those who are disabled, old, or sick, as well as for young children. The emphasis of the healthcare system has evolved throughout time due to several emerging beneficial technologies, such as personal digital assistants (PDAs), data mining, the internet of things, metaheuristics, fog computing, and cloud computing.

Metaheuristics are strong technology for tackling several optimization problems in various fields, especially healthcare systems. The primary advantage of metaheuristic algorithms is their ability to find a better solution to a healthcare problem and their ability to consume as little time as possible. In addition, metaheuristics are more flexible compared to several other optimization techniques. These algorithms are not related to a specific optimization problem but could be applied to any optimization problem by making some small adaptations to become suitable to tackle it.

The successful outcome of this book will enable a decision-maker or practitioner to pick a suitable optimization approach when making decisions to schedule patients under crowding environments with minimized human errors.

Application of Advanced Optimization Techniques for Healthcare Analytics

Mohamed Abdel-Basset, Ripon K. Chakrabortty, Reda Mohamed

CRC Press

Taylor & Francis Group

Boca Raton London New York

CRC Press is an imprint of the
Taylor & Francis Group, an **informa** business

MATLAB® is a trademark of The MathWorks, Inc. and is used with permission. The MathWorks does not warrant the accuracy of the text or exercises in this book. This book's use or discussion of MATLAB® software or related products does not constitute endorsement or sponsorship by The MathWorks of a particular pedagogical approach or particular use of the MATLAB® software.

First edition published 2023
by CRC Press
6000 Broken Sound Parkway NW, Suite 300, Boca Raton, FL 33487-2742

and by CRC Press
4 Park Square, Milton Park, Abingdon, Oxon, OX14 4RN

CRC Press is an imprint of Taylor & Francis Group, LLC

ISBN: 978-1-032-34881-0 (hbk)
ISBN: 978-1-032-35158-2 (pbk)
ISBN: 978-1-003-32557-4 (ebk)

DOI: 10.1201/9781003325574

Typeset in Times
by Newgen Publishing UK

Dedication

Dedicated to our families, who had the patience to allow us to work, even at midnight.

Contents

Preface

1 INTRODUCTION

Many different optimization techniques and methods have been applied in healthcare, ranging from the decision of the operational levels to the design of national healthcare policies. Healthcare facility location, capacity planning, disease screening, and medical human resource scheduling are a few examples, while different optimization techniques and decision analytics (i.e., predictive analytics, prescriptive models) are being successfully used. Besides the traditional optimization concepts, different advanced deep learning (DL) and machine learning (ML) models have also grasped good perspectives in healthcare analytics. For instance, the application of advanced ML or DL models for medical imaging, biomedical signal processing, and DNA microarray data analysis are a few examples.

While the contributions of existing works of literature and their influence on hospital analytics are undeniable, the necessity of better ML or DL approaches, advanced augmentation of ML/DL approaches with metaheuristics, better design of classical optimization tools (e.g., bi-level optimization, nested optimization) are also irrefutable. In addition to the need for advanced solution approaches, the design and articulation of healthcare problems (e.g., home healthcare scheduling, biological sample transportation network, facility location problem, nutrition decision support system, nurse scheduling, smart healthcare scheduling problems) are also evident, particularly during such a tiring era (post-pandemic). Therefore, a combination of better problem architecture and advanced solution approaches to overarching hospital analytics is a must and should be considered for practitioners, undergraduate and postgraduate students, and, most importantly, for higher degree research students.

2 CONTENTS

Considering all these, this book will bring all them together for the first time in academia, covering advanced heuristic, metaheuristic, DL/ML, and prescriptive models, and also encompass different nurse scheduling and medical imaging issues related to hospital analytics. The primary aim of this book is to demonstrate how optimization techniques can be better employed against different hospital analytics problems.

This book consists of 11 chapters. A brief description of the chapters is as follows:

CHAPTER 1 ADVANCED OPTIMIZATION TECHNIQUES

This chapter presents the classification of the optimization algorithms and discusses four different types of modern metaheuristics. The metaheuristics role for healthcare is also reviewed. Finally, the mathematical models of some recently published and highly cited metaheuristic algorithms investigated throughout the book for healthcare optimization problems are briefly described.

Chapter 2 Metaheuristic Algorithms for Healthcare: Open Issues and Challenges

The healthcare system is currently a topic of significant investigation to make life easier for those who are disabled, old, or sick, as well as for young children. The emphasis of the healthcare system has evolved throughout time due to several emerging beneficial technologies, such as personal digital assistants (PDAs), data mining, the internet of things, metaheuristics, fog computing, and cloud computing. This chapter explains different metaheuristic algorithms and their opportunities to apply in the healthcare systems.

Chapter 3 Metaheuristic-Based Augmented Multilayer Perceptrons for Cancer and Heart Disease Predictions

This chapter discusses, at the outset, the feed-forward neural network (FNNs) and multilayer perceptron (MLP). Then, the adaptation of metaheuristic algorithms to train this neural network is discussed. Lastly, the experimental findings, based on various performance metrics, are presented to illustrate the effectiveness of six investigated metaheuristics.

Chapter 4 The Role of Metaheuristics in Multilevel Thresholding Image Segmentation

This chapter discusses Kapur's entropy fitness function employed with various metaheuristic algorithms (MHAs) for segmenting some chest X-ray images under various threshold levels. In addition, the experimental settings and chest X-ray test images are also described. Last but not least, results and a discussion are also presented to illustrate the influence of MHAs on tackling this problem.

Chapter 5 Role of Advanced Metaheuristics for DNA Fragment Assembly Problem

This chapter begins with a discussion of how to design metaheuristic algorithms to solve this problem generally. The choice of control parameters and the experimental settings to solve DFAP are explained. Last, but certainly not least, a few experiments are carried out to determine which algorithm would be the most effective in solving this problem.

Chapter 6 Contribution of Metaheuristic Approaches for Feature Selection Techniques

This chapter, at the start, reviews some previously proposed feature selection techniques and describes how to adapt the metaheuristic algorithms for tackling this problem. Then, the performance metrics and experimental settings are presented. Last but not least, some experiments are conducted to show which algorithm could tackle this problem better.

Chapter 7 Advanced Metaheuristics for Task Scheduling in Healthcare IoT

This chapter investigates the performance of six metaheuristic optimization algorithms for finding the near-optimal scheduling of tasks to the virtual machines (VMs) in fog computing to improve the quality of service presented to the patients in the healthcare system. Various performance metrics: make-span, energy consumption, flow-time, and carbon dioxide emission rate, are considered in this chapter to observe the performance of each metaheuristic algorithm.

Chapter 8 Metaheuristics for Augmenting Machine Learning Models to Process Healthcare Data

At the beginning of this chapter, the support vector machine is carefully overviewed, along with a discussion of how to apply metaheuristic algorithms to better tune the control parameters. The presentation of the choice of regulating parameters and the experimental conditions follows. In conclusion, a few experiments are carried out to identify which algorithm would be the most successful in terms of finding a solution to this problem. Finally, a conclusion of the chapter is provided for your perusal.

Chapter 9 Deep Learning Models to Process Healthcare Data: Introduction

This chapter illustrates the role of metaheuristic algorithms for estimating the hyperparameters of DL techniques to improve their performance accuracy when tackling healthcare tasks. This chapter begins with a discussion of well-known DL techniques like a recurrent neural network, convolutional neural network, and deep neural network to show their hyperparameters, which the metaheuristics could estimate to improve their performance. After that, some well-known optimizers employed to update the weights of the DL models are discussed, in addition to describing some of the activation functions. Next, DL techniques' applications to the healthcare system are discussed. Last but not least, the role of metaheuristics for tuning the hyperparameters of the DL techniques is presented.

Chapter 10 Metaheuristics to Augment DL Models Applied for Healthcare System

This chapter analyses the efficacy of seven alternative metaheuristic methods for optimizing the hyperparameters and structural design of DNN in the interest of detecting COVID-19 disease from chest X-ray images. To be more specific, a group of seven metaheuristic algorithms called GWO, DE, WOA, MPA, SMA, TLBO, and EO are utilized to estimate the number of layers, the number of neurons that make up each layer, the most effective learning, and momentum rate, as well as the most effective optimizer and activation function of DNN to improve its performance when it comes to detecting COVID-19 disease. This was done to enhance the classification performance of the metaheuristic-based DNN models, namely GWO DNN, DE DNN, WOA DNN, MPA DNN, SMA DNN, TLBO DNN, and EO DNN, for detecting COVID-19 disease from chest X-ray images.

CHAPTER 11 INTRUSION DETECTION SYSTEM FOR HEALTHCARE SYSTEM USING DEEP LEARNING AND METAHEURISTICS

The feature scaling methods are presented at the beginning of this chapter. Then, the one-hot-encoding strategy is described. Afterwards, the implementation of the DL model trained using the Keras library's metaheuristic algorithms is discussed. Last but not least, experimental results based on various performance indicators are presented to demonstrate the efficacy of seven investigated metaheuristics.

About the Authors

Mohamed Abdel-Basset (Senior Member, IEEE) received BSc, MSc, and PhD degrees in operations research from the Faculty of Computers and Informatics, Zagazig University, Egypt. He is currently a Professor with the Faculty of Computers and Informatics, Zagazig University. He has published more than 200 articles in international journals and conference proceedings. He is working on the application of multi-objective and robust metaheuristic optimization techniques. His current research interests include optimization, operations research, data mining, computational intelligence, applied statistics, decision support systems, robust optimization, engineering optimization, multi-objective optimization, swarm intelligence, evolutionary algorithms, and artificial neural networks. He is an editor and a reviewer of different international journals and conferences.

Ripon K. Chakrabortty (Senior Member, IEEE) is the Program Coordinator for Master of Project Management, Master of Decision Analytics and Master of Engineering Science Programs, and the team leader of 'Decision Support & Analytics Research Group' at the School of Engineering & Information Technology, UNW Canberra, Australia. He obtained his PhD from the same university in 2017 while completing his MSc and BSc from Bangladesh University of Engineering & Technology in Industrial & Production Engineering in 2013 and 2009, respectively. He has written four book chapters and over 170 technical journal and conference papers. His research interest covers a wide range of topics in operations research, project management, supply chain management, artificial intelligence in decision making, healthcare analytics, and information systems management. Many organizations have funded his research program, such as the Department of Defence, Commonwealth Government, Australia. He is an associate editor and a reviewer of different international journals and conferences.

Reda Mohamed received his BSc degree from the Department of Computer Science, Faculty of Computers and Informatics, Zagazig University, Egypt. He is working on the application of multi-objective and robust metaheuristic optimization techniques in computational intelligence. His research interests include robust optimization, multi-objective optimization, swarm intelligence, evolutionary algorithms, and artificial neural networks.

1 Advanced Optimization Techniques

Introduction

1.1 METAHEURISTIC OPTIMIZATION ALGORITHMS

The application of optimization approaches is eminent in a wide range of fields, including engineering, business, healthcare, and manufacturing. The goals of optimization can be anything—reducing energy consumption and costs, maximizing profit, output, performance, and efficiency. Optimization is necessary for every process, from engineering design to business planning, and Internet routing to vacation planning. The optimization problem can be defined as a computational situation in which the goal is to identify the best feasible solution out of many viable solutions. The optimization problems could be categorized into continuous or discrete, limited or unconstrained, mono or multi-objective, static or dynamic, and many more variations on those basic characteristics. Among the optimization problems, problems dubbed as hard optimization problems cannot be solved to optimality or to any guaranteed bound by any deterministic/exact approach within a "realistic" time constraint. Therefore, metaheuristics can be utilized to develop satisfactory solutions to these issues. An algorithm known as a metaheuristic is one that may be used for a wide variety of difficult optimization problems without the need for extensive adaptation to each one. It is clear from the name that these algorithms are referred to as "higher level" heuristics instead of problem-specific heuristics.

This chapter presents the classification of the optimization algorithms and discusses four different types of modern metaheuristics. The metaheuristics role for healthcare is also reviewed. Finally, the mathematical models of some recently published and highly cited metaheuristic algorithms investigated throughout the book for the healthcare optimization problems will be briefly described.

1.1.1 CLASSIFICATION OF OPTIMIZATION ALGORITHMS

There are numerous methods for classifying an optimization algorithm. Considering the nature of the problem, optimization algorithms can be categorized into two groups: deterministic and non-deterministic. A deterministic algorithm is one in which the output is entirely determined by the model's inputs, where there is no randomness in the model. Given the same inputs, the deterministic algorithm will always return the same outcomes in each operation. This restriction on the outcome of each operation can be removed theoretically, allowing for greater flexibility in operation design. The ability to include operations in algorithms whose outputs are not uniquely determined but are instead restricted to specific groups of possibilities can be achieved. Following a determination condition that will be stated later, the machine that executes each operation will be permitted to choose any possible outcomes. As a result of this, the concept of non-deterministic algorithms is introduced. A non-deterministic algorithm, as an alternative to the deterministic algorithm, always employs randomness to overcome problems with a non-deterministic polynomial-time difficulty.

A non-deterministic polynomial-time algorithm can execute in polynomial or exponential time, depending on the choices made during execution. Non-deterministic algorithms are advantageous for obtaining approximate solutions when a precise solution using a deterministic approach is prohibitively difficult or expensive. While a deterministic algorithm will always produce the same output regardless of how many times it is performed, a non-deterministic algorithm will take alternative paths, allowing it to reach multiple distinct outcomes. Non-deterministic models are typically employed when the problem being solved naturally allows for numerous solutions or when a single outcome may be found by following multiple paths, each of which is equally preferable. In essence, every outcome produced by the non-deterministic algorithm is valid. These conclusions are valid regardless of the algorithm's choices and judgements while running and attempting to solve the problem.

Several deterministic algorithms are proposed for overcoming various optimization problems; some are the trajectory base algorithm, dynamic programming, local search, hill-climbing, and visibility graph search algorithm. Among the non-deterministic algorithms, there is a well-known type known as stochastic algorithms because they always apply randomness to reach the near-optimal solution of the optimization problems. Stochastic algorithms are widely used due to their ability to overcome several optimization problems, especially NP-hard problems. Metaheuristic algorithms are an example of stochastic algorithms. There are two types of metaheuristic algorithms: single- and population-based. Single-based metaheuristic algorithms begin with a single solution and attempt to improve it through several iterations. Some of the most well-known algorithms in this class include Tabu Search [1], Simulated Annealing [2], Variable Neighbourhood Search [3], Hill Climbing [4], and Iterated Local Search [5]. In contrast to single-based metaheuristics, population-based metaheuristics begin with a collection of individuals (population). Then, in each iteration through the optimization process, the current individuals will be updated to explore more regions within the search space. This allows for the flow of information among the collection of solutions. The primary advantage of population-based metaheuristics is its ability to avoid becoming trapped in local optima. This sort of metaheuristic algorithm is a well-known optimization technique that has been widely employed in a variety of applications, including medical systems [6], fault identification [6], car engine design [6], and food quality [6].

Metaheuristic algorithms based on populations have two traits in common: exploitation (intensification) and diversification (exploration). Exploitation is a search strategy that focuses on finding the best possible solutions. When we talk about

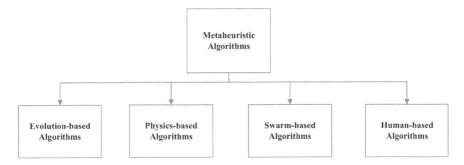

FIGURE 1.1 Classification of metaheuristic algorithms.

exploration, we're talking about the algorithm's ability to look into new areas of potential search space. These two operators are frequently at odds with one another, and determining a reasonable trade-off between intensification and diversification is one of the most difficult problems to solve in developing metaheuristic algorithmic strategies. In general, metaheuristics can be classified into four types: evolution-based algorithms, physics-based algorithms, swarm-based algorithms, and human-based algorithms, as depicted in Figure 1.1; each of which will be discussed in detail within the next sections. It has to be mentioned that no metaheuristic algorithm can optimally handle all optimization problems according to the no free lunch (NFL) theorem [7]. In other words, a metaheuristic algorithm may be extremely efficient in some situations and poorly efficient in others [6].

1.1.2 EVOLUTION-BASED ALGORITHMS

The principles of natural evolution inspire evolution-based algorithms. The search process of these approaches begins with a randomly created population that evolves over several generations. The strength of these approaches is that they always unite the fittest individuals to generate the next generation of individuals. They make use of biologically inspired operators such as crossover and mutation. The most common evolution-based algorithm is the genetic algorithm (GA), which is based on Darwinian evolutionary theory [8]. GA employs the concept of a crossover to generate superior solutions, dubbed offspring, from some suited solutions, dubbed parents. Crossover, which happens in nature, helps preserve the population diversity for exploring the search space of the optimization problems. Because of mutations, the traits of the offspring are different from those of their parents. This GA operator is intended to exploit the neighbouring regions to the individuals. To prevent randomization search, certain solutions and their dimensions undergo mutation, which is described by a function and controlled by a parameter such as mutation probability and percentage. Some of other popular evolution-based algorithms are evolution strategy (ES) [9], probability-based incremental learning (PBIL) [9], genetic programming (GP) [9], differential evolution (DE) [10], and biogeography-based optimizer (BBO) [9].

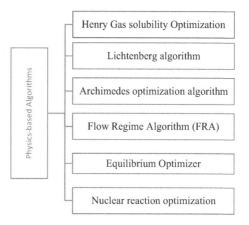

FIGURE 1.2   Some of the recently-published physics-based metaheuristic algorithms.

1.1.3   PHYSICS-BASED ALGORITHMS

These algorithms are derived from natural physical principles and often describe search agents' interaction in governing rules derived from physical processes. Simulated annealing (SA), one of the most popular physics-based algorithms, was designed based on simulating thermodynamics laws applied to heating and then controlled cooling of a material to grow the size of its crystals [9]. The second well-known algorithm in this class is the Gravitational Search Algorithm (GSA), which was based on the law of gravity and mass interactions to update the position toward the optimal point [9]. There are several other physics-based algorithms like chemical reaction optimization (CRO) [9], multiverse optimizer (MVO) [11], lightning search algorithm (LSA) [11], charged system search (CSS) algorithm [11], ray optimization (RO) algorithm [9], colliding bodies optimization (CBO) [12], Big Bang-Big crunch (BB-BC) algorithm [13], vibrating particles system (VPS) algorithm [13], atom search optimization (ASO) [11], equilibrium optimizer (EO) [13], Henry Gas solubility optimization (HGO) [11], artificial physics algorithm (APA) [14], galaxy-based search algorithm (GbSA) [9], black hole (BH) algorithm [15], curved space optimization (CSO) [9], central force optimization (CFO) [9], sine cosine algorithm (SCA) [11], water cycle algorithm (WCA) [13], gravitational local search algorithm (GLSA) [5], small-world optimization algorithm (SWOA) [9], intelligent water drops (IWD) algorithm [13], and Vortex Search Algorithm (VSA) [16]. Figure 1.2 shows some of the recently published physics-based metaheuristic algorithms.

1.1.4   SWARM-BASED ALGORITHMS

These algorithms mimic the collective behaviour of animals (i.e., flocks, herds, or schools). The primary characteristic of these algorithms is that all members share collective information during the optimization process. The most common swarm-based algorithm is particle swarm optimization, developed by Kennedy and Eberhart

[17]. Some of the other swarm-based algorithms are cuckoo search (CS) algorithm [9], flower pollination algorithm (FPA) [9], Harris hawks optimization (HHO) [11], marine predators algorithm (MPA) [11], salp swarm algorithm (SSA) [11], chameleon swarm algorithm [18], artificial gorilla troops optimizer [19], krill herd algorithm [11], donkey and smuggler optimization algorithm [20], elephant herding optimization [21], hunting search [9], wolf pack search algorithm [22], chicken swarm optimization [23], monkey search [24], horse herd optimization algorithm (HOA) [15], earthworm optimization algorithm (EWA) [21], moth search (MS) algorithm [21], slime mould algorithm (SMA) [25], monarch butterfly optimization (MBO) [21] and whale optimization algorithm (WOA) [9].

1.1.5 HUMAN-BASED ALGORITHMS

The human-based algorithms are inspired by emulating human behaviours and interactions in social situations. There are several algorithms belonging to these categories, some of which are teaching-learning-based optimization (TLBO) [9], past present future (PPF) [11], harmony search (HS) [9], political optimizer (PO) [11], exchange market algorithm (EMA) [11], brain storm optimization (BSO) [26], league championship algorithm (LCA) [9], driving training-based optimization [27], poor and rich optimization algorithm [28], imperialist competitive algorithm (ICA) [11], gaining-sharing knowledge-based algorithm (GSK) [29], and soccer league competition (SLC) [9].

1.2 INVESTIGATED METAHEURISTIC ALGORITHMS

Specifically, this book will check the role of metaheuristics in tackling some healthcare problems, such as DNA fragment assembly problems, image segmentation, nurse rostering problems, and others. Seven metaheuristic algorithms selected based on two highly cited and recently published factors will be investigated to disclose their influence on tackling healthcare optimization problems. Those algorithms are marine predators algorithm, equilibrium optimizer, differential evolution, teaching-learning-based optimization, slime mould algorithm, grey wolf optimizer, and whale optimization algorithms. The mathematical model and pseudocode of those algorithms will be discussed in the next subsections.

1.2.1 WHALE OPTIMIZATION ALGORITHM (WOA)

Mirjalili and Lewis [9] used WOA to replicate the humpback whales' emotions and behaviours. When assaulting their victim or prey, the whales encircle them in a spiral pattern and swim up to the surface in a diminishing circle, employing an incredible feeding technique dubbed the bubble-net approach. WOA simulates this hunting mechanism by generating the new position of the current whale with a 50% probability of choosing between a spiral model and a shrinking surrounding prey. To switch between the spiral model and the shrinking encircling mechanism, a random number between 0 and 1 is formed, and if it is less than 0.5, the encircling mechanism is used; otherwise, the spiral model is used. The encircling mechanism's mathematical formula is as follows:

$$\overrightarrow{X_i}(t+1) = \overrightarrow{X^*}(t) - \vec{A} * \vec{D} \tag{1.1}$$

$$\vec{A} = 2 * a * rand - a \tag{1.2}$$

$$a = 2 - 2 * \frac{t}{T} \tag{1.3}$$

$$\vec{D} = \left| \vec{C} * \overrightarrow{X^*}(t) - \overrightarrow{X_i}(t) \right| \tag{1.4}$$

$$C = 2 * rand \tag{1.5}$$

where $\overrightarrow{X_i}$ indicates the current location of the ith whale, t indicates the current iteration, $\overrightarrow{X^*}$ stands for the best-so-far solution, $rand$ is a number generated randomly between 0 and 1, T refers to the maximum iteration, and a is a linearly decreasing distance control parameter from 2 to 0. The spiral model attempts to replicate the helix-shaped movement of whales, and it is proposed between the positions of the victim and that of the whale. The spiral shape is represented mathematically by the following equation:

$$\overrightarrow{X_i}(t+1) = \overrightarrow{X^*}(t) + \cos(2 * \pi * l) * e^{l*b} * \overrightarrow{D'} \tag{1.6}$$

$$\overrightarrow{D'} = \left| \overrightarrow{X^*}(t) - \overrightarrow{X_i}(t) \right| \tag{1.7}$$

where l is a number created randomly between 1 and -1, and b is a constant that describes the logarithmic spiral shape. WOA employs a whale selected randomly from the population to update the position of the current whale in the exploration phase in order to seek prey in a different direction in the search region. If $\vec{A}$ is greater than 1, then the current whale is updated in accordance with the randomly-selected whale from the population. The mathematical model of the search for the prey is defined as follows:

$$\overrightarrow{X_i}(it+1) = \overrightarrow{X_r}(t) - \vec{A} * \vec{D} \tag{1.8}$$

$$\vec{D} = \left| \vec{C} * \overrightarrow{X_r}(t) - \overrightarrow{X_i}(t) \right| \tag{1.9}$$

where $\overrightarrow{X_r}$ is a random whale selected from the population at the current iteration. The pseudocode of WOA is listed in Algorithm 1.1.

Algorithm 1.1 The standard WOA

Output: $\overrightarrow{X^*}$
1. Initialize the population of whales $\overrightarrow{X_i}(i = 1,2,3,.....,N)$
2. Evaluate the fitness of each whale
3. Find the best whale $\overrightarrow{X^*}$
4. $t = 1$
5. **while** ($t < T$)
6.   **for** each i whale
7.   Update a, A, p, C, and l
8.   **if** ($p < 0.5$)
9.   **if** ($|A| < 1$)
10.   Update $\overrightarrow{X_i}(t+1)$ using Eq. (1.1)
11.   **else**
12.   Update $\overrightarrow{X_i}(t+1)$ using Eq. (1.8)
13.   **end if**
14.   **else**
15.   Update $\overrightarrow{X_i}(t+1)$ using Eq. (1.6)
16.   **end if**
17.   Check the objective value of the whale $\overrightarrow{X_i}(t+1)$
18.   Replacing the best whale $\overrightarrow{X^*}$ with $\overrightarrow{X_i}(t+1)$ if better.
19.   **end for**
20. t ++
21. **end while**

1.2.2 Teaching-Learning-Based Optimization

Hashim [30] suggested a population-based technique called teaching-learning-based optimization (TLBO) for addressing the global optimization problem. TLBO was proposed specifically to imitate a teacher's influence on the level of a class's learners. The TLBO is divided into two phases: teacher and learner.

1.2.2.1 Teacher Phase

In this phase, the teacher chosen as the best individual among the learners will attempt to increase the mean value of the learners through knowledge deployment. The mean value of the class's learners is computed as follows:

$$\vec{X}_{mean} = \frac{1}{N}\sum_{i=1}^{N}\vec{X}_i \qquad (1.10)$$

where N indicates the population size, and $\vec{X}_i$ includes the position of the ith solution at the current iteration, t. Finally, each learner in this phase will be updated according to the following formula:

$$\vec{X}_i(t+1) = \vec{X}_i(t) + \vec{r}.\left(\overrightarrow{X^*} - T_F.\vec{X}_{mean}\right) \qquad (1.11)$$

where $\overrightarrow{X^*}$ stands for the teacher vector, $\vec{r}$ stands for a vector assigned randomly within 0 and 1, and T_F may randomly include a value of 1 or 2 to express the teaching factor. The objective value of $\vec{X}_i$ (t+1) is computed, and if it is less than the objective value of $\vec{X}_i(t)$, then $\vec{X}_i(t) = \vec{X}_i(t+1)$.

1.2.2.2 Learner Phase

Discussions among students will aid them in increasing their understanding of the subject matter. During this phase, each learner will select a learner from the population with whom they will engage in a discussion in order to better their overall understanding. This phase is mathematically represented by the following:

$$\vec{X}_i(t+1) = \begin{cases} \vec{X}_i(t) + \vec{r}.\left(\vec{X}_i(t) - \vec{X}_j(t)\right) & \text{if } f\left(\vec{X}_i(t)\right) < f\left(\vec{X}_j(t)\right) \\ \vec{X}_i(t) + \vec{r}.\left(\vec{X}_j(t) - \vec{X}_i(t)\right) & \text{otherwise} \end{cases} \quad (1.12)$$

where $f\left(\vec{X}_i(t)\right)$ and $f\left(\vec{X}_j(t)\right)$ indicates the objective values of the ith and jth learners, respectively. j stands for the index of another learner selected randomly from the current learners, such that $i \neq j$. The objective value of $\vec{X}_i$ (t+1) is computed and if it is less than the objective value of $\vec{X}_i(t)$, then $\vec{X}_i(t) = \vec{X}_i(t+1)$. Finally, the pseudo-code of TLBO is listed in Algorithm 1.2.

Algorithm 1.2 TLBO

    Output: $\overrightarrow{X^*}$
1.   Initialize the population $\vec{X}_i, (i = 1,2,3,\ldots\ldots N)$
2.   Calculate the objective value for each $\vec{X}_i$.
3.   Set the best learner within the population to $\overrightarrow{X^*}$ as the teacher
4. $t = 0$
5.   **While** $(t < T)$
6.   Compute $\vec{X}_{\text{mean}}$ and their objective value.
7. $t = t + 1;$
8.   ////teaching phase
9.   **for** each i solution
10.   Compute $\vec{X}_i(t+1)$ using Eq. (1.11)
11.   If $f\left(\vec{X}_i(t+1)\right) < f\left(\vec{X}_i(t)\right)$
12. $\vec{X}_i(t) = \vec{X}_i(t+1)$
13.   End if
14.   **End for**
15. $t = t+1$
16.   //learning phase
17.   **for** each i solution
18.   Compute $\vec{X}_i(t+1)$ using Eq. (1.12)
19.   If $f\left(\vec{X}_i(t+1)\right) < f\left(\vec{X}_i(t)\right)$
20. $\vec{X}_i(t) = \vec{X}_i(t+1)$

21.   End if
22.   **End for**
23.   t = t+1
24.   **End while**

___

1.2.3 EQUILIBRIUM OPTIMIZER

Equilibrium optimizer (EO), a novel physics-based metaheuristic algorithm for tackling optimization issues, particularly single continuous optimization problems, has recently been described [31]. It has the potential to outperform numerous competing algorithms. EO was developed to simulate a general mass balance equation for a controlled volume and aims to identify the equilibrium state that is true to the balancing between the quantity of mass that enters, exits, and is found in the system. The mathematical model of EO is discussed in greater depth later in this section.

Before beginning the optimization process, by Equation 1.13, EO generates N solutions that are distributed throughout the search space of an optimization problem that has d-dimensions/variables and evaluates each one using an objective function, which may be maximized or minimized depending on the nature of the optimization problem being solved. Following that, it identifies the best four solutions and assigns them to a vector, namely $\vec{C}_{eq,pool}$, in addition to computing the average of those best solutions and storing it in the same vector, as shown in Equation 1.14.

$$\vec{X}_i = \vec{X}_L + \left(\vec{X}_U - \vec{X}_L\right) * \vec{r} \quad i = 0,1,2,\ldots,N \tag{1.13}$$

$$\vec{C}_{eq,pool} = \left[\vec{X}_{eq(1)}, \quad \vec{X}_{eq(2)}, \quad \vec{X}_{eq(3)}, \quad \vec{X}_{eq(4)}, \quad \vec{X}_{eq(avg)}\right] \tag{1.14}$$

where, $\vec{X}_L$ and $\vec{X}_U$ includes the values for the lower bound and upper bound limits of the dimensions, and $\vec{r}$ is a vector including values generated randomly between 0 and 1.

To begin the optimization process, EO uses the following equation to calculate the value of a term, known as t_1:

$$t_1 = \left(1 - \frac{t}{T}\right)^{\left(a_2 * \left(\frac{t}{T}\right)\right)} \tag{1.15}$$

In this case, a_2 is a constant that is used to govern the exploitation operator of EO. An exponential term (F) is related to this term, and it is intended to strike a balance between exploration and exploitation capabilities. The mathematical representation of the term F is as follows [31]:

$$\vec{F} = a_1 * sign\left(\vec{r} - 0.5\right) * [e^{-\vec{\lambda}(t_1)} - 1] \tag{1.16}$$

where $\vec{r}$ and $\vec{\lambda}$ are two random vectors that are generated at the range of 0 and 1, and a_1 is a fixed value that controls the exploration capability. In addition, a new factor called the generation rate ($\vec{G}$) has been used by EO to even more improve its exploitation operator. $\vec{G}$ is mathematically defined as that:

$$\vec{G} = \vec{G}_0 * \vec{F} \tag{1.17}$$

where $\vec{G}_0$ is described as follows:

$$\vec{G}_0 = \overrightarrow{GCP} * \left(\overrightarrow{c_{eq}} - \vec{\lambda} * \overrightarrow{X_i} \right) \tag{1.18}$$

$$\overrightarrow{GCP} = \begin{cases} 0.5r_1 & r_2 \geq GP \\ \\ 0 & otherwise \end{cases} \tag{1.19}$$

where r_1 and r_2 are random numbers between 0 and 1, and $\overrightarrow{(GCP)}$ is a vector containing values between 0 and 1 that indicates when the generation rate will be applied to the optimization process based on a specified generation probability (GP), recommended at 0.5 in the original study. $\overrightarrow{(c_{eq})}$ is a solution chosen at random from the five solutions found in $\vec{C}_{eq,pool}$, and $\overrightarrow{(X_i)}$ represents the ith solution's current position. Finally, The following equation will be used to update each solution during the optimization phase:

$$\overrightarrow{X_i} = \overrightarrow{c_{eq}} + \left(\overrightarrow{X_i} - \overrightarrow{c_{eq}} \right) * \vec{F} + \frac{\vec{G}}{\vec{\lambda} * V} * \left(1 - \vec{F} \right) \tag{1.20}$$

where V is the velocity value, and the suggested value is 1. The pseudocode of EO is described in Algorithm 1.3.

Algorithm 1.3 EO

---

1. Initialization.
2. while $(t < T)$
3.   Compute the objective value of the solutions $f(\vec{X})$
4. $\vec{C}_{avg} = \left(\vec{X}_{eq(1)} + \vec{X}_{eq(2)} + \vec{X}_{eq(3)} + \vec{X}_{eq(4)} \right) / 4$
5. $\vec{C}_{eq,pool} = \left[\vec{X}_{eq(1)}, \vec{X}_{eq(2)}, \vec{X}_{eq(3)}, \vec{X}_{eq(4)}, \vec{X}_{eq(avg)} \right]$
6.   Apply the memory saving mechanism.
7.   Compute t_1 using (1.15)
8.   **for** each i solution
9.   select a solution randomly from $\vec{C}_{eq,pool}$
10.   compute $\vec{F}$ using (1.16)
11.   compute $\overrightarrow{GCP}$ using (1.19)
12.   compute $\vec{G}_0$ using (1.18)
13.   compute $\vec{G}$ using (1.17)
14.   Update each $\vec{X}_i$ using (1.20)
15.   **end for**
16. $t++$
17. **End while**

1.2.4 GREY WOLF OPTIMIZER

A unique metaheuristic algorithm dubbed the grey wolf optimizer (GWO) was introduced and inspired by the behaviour of grey wolves when hunting for, encircling, and catching their prey [32]. Grey wolves are classified into four types in GWO based on their dominance and leadership: alpha α, beta β, delta δ, and omega ω. α is considered the best solution so far, β is the second best solution, δ is the third-best solution, and ω represents the rest of the wolves. Using the following formula, the wolves can mathematically encircle their victim during the course of their hunting expedition:

$$\vec{D} = \left| \vec{C}.\vec{X}_p - \vec{X}(t) \right| \tag{1.21}$$

$$\vec{X}(t+1) = \vec{X}_p(t) - \vec{D}.\vec{A} \tag{1.22}$$

where $\vec{X}(t)$, and $\vec{X}_p$ correspond to the current iteration's position of the prey and grey wolf, respectively. $\vec{D}$ is a vector containing the absolute value of the difference between the prey vector $\vec{X}_p$ multiplied by the coefficient vector $\vec{C}$ in order to avoid local minima and the grey wolf vector $\vec{X}(t)$. $\vec{C}$ is generated according to the following formula:

$$\vec{C} = 2 * \vec{r_1} \tag{1.23}$$

where $\vec{r_1}$ is a randomly generated vector ranging from 0 to 1. In relation to the coefficient vector $\vec{A}$, it is a factor that influences the exploration and exploitation capability of the optimization process and is expressed as:

$$\vec{A} = 2 * \vec{a} * \vec{r_2} - \vec{a} \tag{1.24}$$

$\vec{r_2}$ is a vector assigned random values ranging from 0 to 1. $\vec{a}$ is a distance control factor that begins with a value equal to 2 and gradually decreases linearly until it reaches 0. Generally, $\vec{a}$ could be computed based on the following formula:

$$\vec{a} = 2 - 2 * \frac{t}{T} \tag{1.25}$$

The preceding mathematical model simulates the grey wolves' encircling behaviour. GWO proposed that the top three grey wolves thus far (α, β, δ, and ω) be aware of the possible prey's location and that the other grey wolves change their positions in accordance with the top three grey wolves. The grey wolf's hunting method can be expressed mathematically as follows:

$$\vec{X}(t+1) = \left(\vec{X_1} + \vec{X_2} + \vec{X_2} \right) / 3.0 \tag{1.26}$$

$$\overrightarrow{X_1} = \overrightarrow{X_\alpha} - \overrightarrow{A_1}.\overrightarrow{D_\alpha}, \quad \overrightarrow{D_\alpha} = \left| \overrightarrow{C_1}.\overrightarrow{X_\alpha} - \vec{X} \right| \tag{1.27}$$

$$\overrightarrow{X_2} = \overrightarrow{X_\beta} - \overrightarrow{A_2}.\overrightarrow{D_\beta}, \quad \overrightarrow{D_\beta} = \left| \overrightarrow{C_2}.\overrightarrow{X_\beta} - \vec{X} \right| \tag{1.28}$$

$$\overrightarrow{X_3} = \overrightarrow{X_\delta} - \overrightarrow{A_3}.\overrightarrow{D_\delta}, \quad \overrightarrow{D_\delta} = \left| \overrightarrow{C_3}.\overrightarrow{X_\delta} - \vec{X} \right| \tag{1.29}$$

When $\vec{A}$ is between 1 and -1, a grey wolf's next location is anywhere between its current and prey positions. On the contrary, $\vec{A} > 1$ and $\vec{A} < -1$ compels grey wolves to depart from their prey in search of better prey. The pseudocode of GWO is described in Algorithm 1.4.

Algorithm 1.4　Grey wolf optimizer (GWO)

Output: X_α
1.　Initialization step.
2.　initialize a, A, and C.
3.　Compute the fitness value for each grey wolf
4.　Select the best three ones as X_α, X_β, and X_δ, respectively
5.　t=0;
6.　**while** (t < T)
7.　　update each grey wolf using Eq.(1.26)
8.　　Calculate the fitness value for each grey wolf.
9.　　Update a, A, and C
10.　Calculate the fitness of the grey wolf
11.　Update the X_α, X_β, and X_δ
12.　t++
13.　end while

1.2.5　Slime Mould Algorithm

Li et al. [33] have introduced a new optimization technique called the slime mould algorithm (SMA), which mimics the slime mould's feeding mechanism. The following sections outline the basic structure and phases of SMA.

1.2.5.1　Searching for Food

By utilizing the odour in the air, the slime mould is able to reach the food. This can be mathematically represented using the following formula:

$$\vec{X}(t+1) = \begin{cases} \overrightarrow{X^*}(t) + \vec{vb} * \left(\vec{W} * \overrightarrow{X_A}(t) - \overrightarrow{X_B}(t) \right), r < p \\ \vec{vc} * \vec{X}(t), r \geq p \end{cases} \tag{1.30}$$

where $\vec{vb}$ is a vector including values generated randomly between a and $-a$ as follows:

$$\vec{vb} = [-a, a] \tag{1.31}$$

$$a = \text{arctanh}\left(-\left(\frac{t}{t_{max}}\right) + 1\right) \tag{1.32}$$

$\vec{vc}$ is a vector including values decreasing linearly from 1 to 0, and r a number generated randomly between 0 and 1. X_A, and X_B are two solutions selected randomly from the population. p is estimated according to the following formula:

$$p = \tanh\left(|f(i) - bF|\right) \tag{1.33}$$

where $f(i)$ is the fitness of the ith the solution, and bF stands for the best fitness obtained even now. $\vec{W}$ indicates the weight of the slime mould, which is computed according to the following formula:

$$\vec{W}(smellindex(l)) = \begin{cases} 1 + r * log\left(\dfrac{bF - f(i)}{bF - wF} + 1\right), condition \\ \\ 1 - r * log\left(\dfrac{bF - f(i)}{bF - wF} + 1\right), other \end{cases} \tag{1.34}$$

$$smellindex = sort(f) \tag{1.35}$$

where r is a number generated randomly between 0 and 1, wF stands for the worst fitness value, $smellindex$ indicates the sorted fitness value's indices, $condition$ indicates the first half of the population's rankings.

1.2.5.2 Wrap Food

This phase simulates the contraction mode of the slime mould's venous tissue structure; the slime mould generates new solutions or positions based on the quality of the food as follows: when the concentration of the food is high, the weight of this region is greater; when the concentration of the food is low, the weight of this region is turned to explore other regions, as shown in Equation (1.34). The slime mould must decide when to leave the present location for another until it can locate multiple food sources concurrently with the current one. In general, the mathematical model for updating the slime mould's position might be rewritten, as stated in Equation (1.36), to imitate the slime mould's methodology when roaming another area in search of a variety of food sources simultaneously.

$$\vec{S}(t+1) = \begin{cases} \vec{X_L} + \left(\vec{X_U} - \vec{X_L}\right) * \vec{r}, & r_1 < z \\ \vec{X}^*(t) + \vec{vb} * \left(\vec{W} * \vec{X_A}(t) - \vec{X_B}(t)\right), & r < p \\ \vec{vc} * \vec{X}(t), & r \geq p \end{cases} \tag{1.36}$$

where r_1, are a number generated randomly between 0 and 1. z is a likelihood that indicates whether the present mould will seek out another food source to avoid becoming caught in a local optimum or will continue to use the existing one to find better solutions. $\vec{W}$, $\vec{vb}$, and $\vec{vc}$ are used to simulate venous width variation. Finally, Algorithm 1.5 depicts the steps of SMA.

Algorithm 1.5 The SMA

1. Initializations step
2. **While** ($t < T$)
3.   Compute the fitness of each $\vec{X}_i$
4.   Update dF and $\overrightarrow{X^*}$ if there is a better solution
5.   Calculate W using Eq. (1.34)
6.   **for** each X_i mould
7.   Update p, vb, and vc
8.   Update $\vec{X}_i$ using Eq. (1.36)
9.   **end for**
10. $t++$
11. **end while**
12. Output: return DF, $\overrightarrow{S_b}$

1.2.6 DIFFERENTIAL EVOLUTION

Storn [10] introduced a population-based optimization algorithm called differential evolution (DE) that, in terms of mutation, crossover, and selection operators, is similar to genetic algorithms. Before beginning the optimization process, the differential evolution creates a set of individuals with D dimensions for each individual, randomly distributed within the search space of the optimization problem. Following then, as stated below, the mutation and crossover operators were used to explore the search space in order to identify better solutions.

1.2.6.1 Mutation Operator

With the use of this operator, DE has been able to construct one mutant vector, $\vec{v}_i^t$, for each individual $\overrightarrow{X_i}$ in the population. Following the mutation process outlined below, the mutant vector is created.

$$\vec{v}_i^t = \overrightarrow{X_a}(t) + F.\left(\overrightarrow{X_k}(t) - \overrightarrow{X_j}(t)\right) \qquad (1.37)$$

where a, k, and j are indices of three solutions selected randomly from the population at generation t. F is a positive scaling factor.

1.2.6.2 Crossover Operator

Following the generation of the mutant vector v_i^t, the crossover operator was used to construct a trial vector $\vec{u}_i^t$ based on the ith individual's present position and the matching mutant one under a crossover probability, namely CR. The description of this crossover operation is as follows:

$$u_{i,j}^t = \begin{cases} v_{i,j}^t & if\left(r_1 \le CR\right) \| \left(j = j_r\right) \\ X_{i,j}\left(t\right) & otherwise \end{cases} \quad (1.38)$$

where j_r is an integer generated randomly between 1 and D, j stands for the current dimension, and CR is a constant value between 0 and 1 that specifies the percentage of dimensions copied from the mutant vector to the trial vector.

1.2.6.3   Selection Operator

Finally, the selection operator is employed to compare the trial vector $\vec{u}_i^t$ it to the current one $\overrightarrow{X_i}(t)$, with the fittest one being used in the subsequent generation. Generally, the selection procedure for a minimization problem is described mathematically as that:

$$\overrightarrow{X_i}(t+1) = \begin{cases} \vec{u}_i^t & if\left(f\left(\vec{u}_i^t\right) < f\left(\overrightarrow{X_i}(t)\right)\right) \\ \overrightarrow{X_i}(t) & otherwise \end{cases} \quad (1.39)$$

1.2.7   MARINE PREDATORS ALGORITHM (MPA)

Faramarzi et al. [34] have developed a novel metaheuristic algorithm, dubbed the marine predators' algorithm (MPA), that replicates predators' behaviour when attacking prey. Specifically, predators make a trade-off between Levy's flight and Brownian strategy in response to the velocity of their prey. MPA is defined as follows:

As is the case with the majority of metaheuristic algorithms, the following equation will be used to disperse N solutions (prey) within the search space at the start of the optimization phase:

$$\vec{A} = \overrightarrow{X_L} + \left(\overrightarrow{X_U} - \overrightarrow{X_L}\right) * \vec{r} \quad (1.40)$$

The fitness function is then determined, and the prey with the highest fitness is chosen as the Top predator throughout the optimization phase to produce the Elite (E) matrix. The following illustrates how this matrix is formed:

$$E = \begin{bmatrix} A_{1,1}^I & A_{1,2}^I & \cdots & A_{1,d}^I \\ A_{2,1}^I & A_{2,2}^I & \cdots & A_{2,d}^I \\ \cdot & \cdot & \cdot & \cdot \\ \cdot & \cdot & \cdot & \cdot \\ A_{N,1}^I & A_{N,2}^I & \cdots & A_{N,d}^I \end{bmatrix}$$

A^I is the top predator and must be repeated N times in order to form the elite matrix. N denotes the population's individual numbers, while d is each individual's dimension number. During the optimization process, the predators are directed towards a prey matrix that has been formulated and initialized randomly throughout the search space:

$$\vec{X} = \begin{bmatrix} A_{1,1} & A_{1,2} & \cdots & A_{1,d} \\ A_{2,1} & A_{2,2} & \cdots & A_{2,d} \\ \cdot & \cdot & \cdot & \cdot \\ \cdot & \cdot & \cdot & \cdot \\ A_{N,1} & A_{N,2} & \cdots & A_{N,d} \end{bmatrix}$$

Through the optimization process, Following the velocity ratio between the prey and the predators, each prey's update will be divided into three stages.

1.2.7.1 High-Velocity Ratio

The velocity ratio between prey and predators is high during this phase, as the prey moves rapidly in search of food and the predators closely monitor their movements. As a result, predators in this phase are not required to move at all, as the prey will reach themselves. This phase happens at the beginning of the optimization process, during which the algorithm searches for a more optimal solution in all regions of the problem's search space. This phase is expressed numerically as follows:

$$while \quad t < \frac{1}{3}*T$$

$$\vec{S}_i = \vec{R}_B \otimes \left(\vec{E}_i - \vec{R}_B \otimes \overrightarrow{X_i} \right) \tag{1.41}$$

$$\vec{X}_i = \vec{X}_i + P*\vec{R} \otimes \vec{S}_i \tag{1.42}$$

Where $\vec{R}_B$ is a random vector allocated to represent the Brownian strategy using the normal distribution, $\otimes$ is the entry-wise multiplication, P a constant value set to 0.5 as specified in the original paper, $\vec{R}$ is a random vector created between 0 and 1.

1.2.7.2 Unit Velocity Ratio

This stage happens during the optimization process's intermediate stage when the exploration phase is transformed into the exploitation phase. As a result, this stage is neither exploration nor exploitation but a combination of the two. As a result, MPA divides the population into two halves during this phase: the first half will proceed through exploration steps, while the second half will move through exploitation steps. Finally, this stage will be expressed mathematically as follows:

$$\text{while} \qquad \frac{1}{3}*T \le t < \frac{2}{3}*T$$

- The first half of the population

$$\vec{S}_i = \vec{R}_L \otimes \left(\vec{E}_i - \vec{R}_L \otimes \overrightarrow{X_i} \right) \qquad\qquad (1.43)$$

$$\vec{X}_i = \vec{X}_i + P * \vec{R} \otimes \vec{S}_i \qquad\qquad (1.44)$$

- The second half of the population

$$\vec{S}_i = \vec{R}_B \otimes \left(\vec{R}_B \otimes \vec{E}_i - \overrightarrow{X_i} \right) \qquad\qquad (1.45)$$

$$\vec{X}_i = \vec{E}_i + P * CF \otimes \vec{S}_i \qquad\qquad (1.46)$$

Where $\vec{R}_L$ is a vector including random values based on the levy distribution, and CF is an adaptive parameter that was built using Equation 1.47 to control the step size:

$$CF = \left(1 - \frac{t}{T} \right)^{\left(2\frac{t}{T} \right)} \qquad\qquad (1.47)$$

1.2.7.3 Low Velocity-Ratio

During this stage, which occurs at the end of the optimization process, the exploration operator is totally transformed into an exploitation operator and mathematically modelled as:

$$\text{while} \qquad t \ge \frac{2}{3}*T$$

$$\vec{S}_i = \vec{R}_L \otimes \left(\vec{R}_L \otimes \vec{E}_i - \overrightarrow{X_i} \right) \qquad\qquad (1.48)$$

$$\vec{X}_i = \vec{E}_i + P * CF \otimes \vec{S}_i \qquad\qquad (1.49)$$

Additionally, eddy formation and FADs (fish aggregating devices) considerably impact predatory behaviour. In some research, predators spend 20% of their time searching for prey outside of the search area, while the rest of their time is spent looking for a better solution in the surrounding environment. The following formula can be used to simulate FADs:

$$\vec{X}_i = \begin{cases} \vec{X}_i + CF * \left[\vec{X}_L + \vec{r}_2 (\vec{X}_U - \vec{X}_L) \right] \otimes \vec{U} & \text{if } r < FADs \\ \vec{X}_i + \left[FADs(1-r) + r \right] \left(\vec{X}_k - \vec{X}_j \right) & \text{if } r \ge FADs \end{cases} \qquad (1.50)$$

where r is a number generated randomly between 0 and 1, $\vec{r_2}$ is a vector including values generated randomly between 0 and 1. $\vec{U}$ is a binary vector containing 0 and 1. FADs $= 0.2$ reflects the likelihood that FADs will have an effect on the optimization process.

MPA compares the fitness of updated solutions to the fitness of the preceding solution following each updating procedure to determine whether the update improved the positions or not. Suppose the updated position of each solution is superior to the old one. In that case, the updated solution will be stored for comparison with the next generation, and if the old solution is superior to the updated one, the old solution will be used in the next generation rather than the updated one. Memory saving is the term used to describe this operation. The pseudocode of MPA is described in Algorithm 1.6.

Algorithm 1.6 The marine predators algorithm (MPA)

1.   Initialize $X_i i = (1,2,3,\ldots\ldots,N)$, $P = 0.5$
2.   **while** $(t < T)$
3.   Calculate the fitness of each $\vec{X}_i$.
4. $\vec{X}^* = $ the fittest solution.
5. $b_{fit} = $ the fitness value of $\vec{X}^*$.
6.   Implement the memory saving
7.   Build E matrix at the first iteration and update it later if there is a better
8.   compute CF according to Eq. (1.47)
9.   **for** each i solution
10.   *if* $(t < \frac{1}{3} * T)$
11.   move the current $\vec{X}_i$ to another position based on Eq. (1.42)
12.   *Else if* $(\frac{1}{3} * T < t < \frac{2}{3} * T)$
13.   *If* $(i < \frac{1}{2} * N)$
14.   move the current $\vec{X}_i$ to another position based on Eq. (1.44)
15.   *Else*
16.   move the current $\vec{X}_i$ to another position based on Eq. (1.46)
17.   *end if*
18.   *Else*
19.   move the current $\vec{X}_i$ to another position based on Eq. (1.49)
20.   *end if*
21.   **end for**
22.   Compute the fitness of each $\vec{X}_i$.
23.   Update E, if there is a newly updated solution better than the current $\vec{X}^*$
24.   implement the memory saving
25.   Execute the FADS according to Eq. (1.50)
26. $t++$
27.   **end while**
28.   **Return** Best$_{fit}$

1.3 IMPLEMENTATION

This section presents the implementation code of WOA in Java programming language as an example to show to code any metaheuristic algorithm. Briefly, the metaheuristic algorithms consist of two consecutive stages: initialization of population and optimization process. Starting with the initialization stage, in this stage, a population of N solutions will be distributed within the search space of the optimization problem as described in the previous section. This stage is coded in Java as shown in Code 1 to optimize a well-known mathematical problem called sphere, which is programmed in Code 2. Code 1 is based on creating a function called Initialization with five parameters: the first parameter is a reference to a two-dimensional array, the second one is a one-dimensional array containing the upper bounds of the optimization problem dimensions, and the last parameter includes the lower bounds. Afterwards, **Line 1** defines a reference variable, namely rand, to contain the address of an object created from the Random class. Following that, the main initialization stage is fired within **Lines 2–6** to initialize the two-dimensional array, X, within the lower and upper bounds of the optimization problem.

Code 1 Initialization stage in Java

```
Public void Initialization (double X[][], double XU [], double XL[])
{
1.   Random rand=new Random(), //Create a random object from the Random class
2.   for (int i=0; i<X.length; i++){
3.     for (int j=0; j<X[0].length; j++){
4.       X[i][j]=XL[j]+ rand.nextDouble()* (XU[j]− XL[j]);
5.     }
6.   }
}
```

Code 2 Fitness function: Sphere in Java

```
Public double fun (double X[])
{
7.   int sum=0;
8.   for (int i=0; i<X.length; i++){
9.     sum=sum+X[i]*X[i];
10.  }
return sum;
}
```

Code 3 shows the steps of the WOA in Java. The first two lines in this code initialize the number of dimensions (*dim*) of the mathematical sphere problem and the population size (*N*). The next line (Line 3) creates an array of N*dim, namely X, to contain the positions of the population within the optimization process. Afterwards, two arrays, namely XU and XL, are created to store the upper and lower bounds of each dimension in the optimization problem, respectively. **Line 10** initializes each dimension in the array X randomly within its upper and lower bounds given in the arrays XL and XU. The current iteration and the maximum number of iterations are defined in **Lines 11** and **12** to determine the end of the optimization process. Lines 16–49 represent the optimization process of WOA, which is repeated more and more gain until the termination condition is satisfied.

Code 3 WOA in Java

```
 1.  int dim=30; Number of dimensions of an optimization problem
 2.  int N=5; Population size
 3.  double X[][]=new double[N][dim];
 4.  int XU[]=new int[dim];
 5.  int XL[]=new int[dim];
 6.  for (int j=0; j<dim; j++){
 7.      XL[j]=-5.12;
 8.      XU[j]=5.12;
 9.  }
10.  Initialization (X, XU, XL);// Initialize N solutions
11.  int t=0;// Create a variable to represent the current iteration
12.  int max_iter=500; // a variable to contain the maximum number of iteration
13.  Random rand=new Random(), //Create a random object from the Random class
14.  double best_Ft=Double.MAX_VALUE; a variable to store the best-so-far fitness for a
     minimization problem
15.  double best_Sol[];
16.  while (t < max _iter){
17.      double a = (3 - 3 * (((double) it / max_iter)));
18.      double r11 = (rand.nextDouble());
19.      double A = a * (2 * r11 - 1);
20.      double C = 2 * r11;
21.      double r4 = (rand.nextDouble());
22.      double t = (rand.nextDouble() * 2 - 1);
23.      // Evaluate the current position using the sphere objective function in Code 2
24.      for (int i=0; i<X.length; i++){
25.          double f=fun (X[i]);// compute the fitness value of the ith solution
26.          if ( f < best _ Ft){
27.              best_Ft=f;
28.              best_Sol= X[i].clone();
29.          }
30.      }
31.      for (int i=0; i<N; i++){
32.          r4 = (rand.nextDouble());
33.          double a2 = -1 + it * (-1.0 / max_iter);
34.          double l = (a2 - 1) * rand.nextDouble() + 1;
35.      for (int j=0; j<dim; j++){
36.          if (r4 <= 0.5) {
37.              if (Math.abs(A) < 1) {
38.                  X[i][j] = best_Sol [j] - A * Math.abs(C * best_Sol[n] - X[i][j]);
39.              } else {
40.                  int x = (rand.nextInt(N));
41.                  X[i][j] = X[i][j] - A * Math.abs(C * X[i][j] - X[i][j]);
```

```
42.              }
43.            } else {
44.                X[i][j] = best_Sol [j] + Math.cos(2 * Math.PI * 1) * Math.exp(b*1) *
                   Math.abs(best_Sol [j] - X[i][j]);
45.            }
46.            // Check the search boundary of the updated jth dimension
47.            if (X[i][j]<XL[j]) {
48.                X[i][j]=XL[j];
49.            }
50.            if (X[i][j]>XU[j]) {
51.                X[i][j]=XU[j];
52.            }
53.          }
54.      }
55.    t++;
56.  }
return best_Ft;
```

1.4   CHAPTER SUMMARY

This chapter presents the classification of the optimization algorithms and discusses
four different types of modern metaheuristics. The evolution-based, swarm-based,
physics-based and human-based algorithms are discussed along with their pseudocode for future implementation. Considering this book's introductory chapter, the
key metaheuristic algorithms used throughout the book are discussed in more detail.
The metaheuristic's role in healthcare is also reviewed. Furthermore, the mathematical models of some recently published and highly cited metaheuristic algorithms
investigated throughout the book for the healthcare optimization problems are briefly
described.

1.5   EXERCISES

a.   What is a metaheuristic algorithm? How is that different from traditional opti-
     mization techniques?
b.   What is a non-deterministic polynomial-time algorithm? Why do we
     need metaheuristics to solve large non-deterministic polynomial-time
     algorithms?
c.   Based on populations, how do you classify metaheuristic algorithms? What
     are they?
d.   Why "a metaheuristic algorithm may be extremely efficient in some situations
     and poorly efficient in others"? Justify your answer.
e.   What is a physics-based metaheuristic algorithm? Name a few and briefly
     explain how they are different from swarm-based algorithms.
f.   Briefly explain the methodology for slime mould algorithm. How is this algo-
     rithm different from the grey wolf optimizer?

REFERENCES

1. Glover, F. Tabu search: part I. *ORSA Journal on Computing*, 1989. **1**(3): pp. 190–206.
2. Hwang, C.-R., Simulated annealing: Theory and applications. *Acta Appl Math*, 1988. 12: 108–111.
3. Hansen, P., N. Mladenović, and J. Moreno Perez, Variable neighbourhood search: Methods and applications. *Annals of Operations Research*, 2008. **6**(4): pp. 319–360.
4. Mitchell, M., J. Holland, and S. Forrest, *When will a genetic algorithm outperform hill climbing*. 1993, Morgan-Kaufmann.
5. Webster, B. and P.J. Bernhard, *A local search optimisation algorithm based on natural principles of gravitation*. 2003, Florida Institute of Technology.
6. Mousavirad, S.J. and H. Ebrahimpour-Komleh, Human mental search: A new population-based metaheuristic optimisation algorithm. *Applied Intelligence*. 2017. **47**(3): pp. 850–887.
7. Wolpert, D.H. and W.G. Macready, No free lunch theorems for optimisation. *IEEE Transactions on Evolutionary Computation*, 1997. **1**(1): pp. 67–82.
8. Holland, J.H., Genetic algorithms. *Scientific American*, 1992. **267**(1): p. 66–73.
9. Mirjalili, S. and A. Lewis, The whale optimisation algorithm. *Advances in Engineering Software*, 2016. **95**: pp. 51–67.
10. Storn, R., International Computer Science Institute, Differential evolution-a simple and efficient adaptive scheme for global optimisation over continuous spaces. *International Computer Science Institute*, 1995. **11**.
11. Naik, A. and S.C., Satapathy, Past present future: A new human-based algorithm for stochastic optimisation. *Soft Computing*, 2021. **25**(20): pp. 12915–12976.
12. Kaveh, A., V.R. Mahdavi, Colliding bodies optimisation: A novel meta-heuristic method. *Computers and Structures*, 2014. **139**: pp. 18–27.
13. Abdollahzadeh, B., et al., African vultures optimisation algorithm: A new nature-inspired metaheuristic algorithm for global optimisation problems. *Computers and Industrial Engineering*. 2021. **158**: p. 107408.
14. Xie, L., J. Zeng, and Z. Cui. General framework of artificial physics optimisation algorithm. In *2009 World Congress on Nature & Biologically Inspired Computing (NaBIC)*. 2009. IEEE.
15. MiarNaeimi, F., G. Azizyan, and M. Rashki, Horse herd optimisation algorithm: A nature-inspired algorithm for high-dimensional optimisation problems. *Knowledge-Based Systems,* 2021. **213**: p. 106711.
16. Doğan, B. and T. Ölmez, A new metaheuristic for numerical function optimisa-tion: Vortex search algorithm. *Information Sciences,* 2015. **293**: pp. 125–145.
17. Kennedy, J. and R. Eberhart. Particle swarm optimisation. In *Proceedings of ICNN'95-international Conference on Neural Networks*. 1995, IEEE.
18. Braik, M.S., Chameleon swarm algorithm: A bio-inspired optimiser for solving engin-eering design problems. *Expert Systems with Applications,* 2021. **174**: p. 114685.
19. Abdollahzadeh, B., F. Soleimanian Gharehchopogh, and S.J. Mirjalili, Artificial gorilla troops optimiser: A new nature-inspired metaheuristic algorithm for global optimisation problems. *International Journal of Intelligent Systems,* 2021. **36**(10): pp. 5887–5958.
20. Shamsaldin, A.S., et al., Donkey and smuggler optimisation algorithm: A collaborative working approach to path finding. *Journal of Computational Design and Engineering,* 2019. **6**(4): pp. 562–583.

21. Wang, G., Moth search algorithm: A bio-inspired metaheuristic algorithm for global optimisation problems. *Memetic Computing,* 2018. **10**(2): pp. 151–164.

22. Yang, C., X. Tu, and J. Chen. Algorithm of marriage in honey bees optimisation based on the wolf pack search. In *The 2007 International Conference on Intelligent Pervasive Computing (IPC 2007)*. 2007, IEEE.

23. Meng, X., et al. A new bio-inspired algorithm: chicken swarm optimisation. In *International Conference in Swarm Intelligence*. 2014, Springer.

24. Mucherino, A. and O. Seref. Monkey search: A novel metaheuristic search for global optimisation. In *AIP Conference Proceedings*. 2007, American Institute of Physics.

25. Li, S., et al., Slime mould algorithm: A new method for stochastic optimisation. *Future Generation Computer Systems,* 2020. **111**: pp. 300–323.

26. Shi, Y. Brain storm optimisation algorithm. In *International Conference in Swarm Intelligence*. 2011. Springer.

27. Dehghani, M., E. Trojovská, and P. Trojovský, Driving training-based optimisation: a new human-based metaheuristic algorithm for solving optimisation problems. *Scientific Reports*, 2022.

28. Moosavi, S.H.S. and V.K. Bardsiri, Poor and rich optimisation algorithm: A new human-based and multi populations algorithm. *Engineering Applications of Artificial Intelligence*, 2019. **86**: pp. 165–181.

29. Mohamed, A.W., et al., Gaining-sharing knowledge based algorithm for solving optimisation problems: A novel nature-inspired algorithm. *International Journal of Machine Learning and Cybernetics,* 2020. **11**(7): pp. 1501–1529.

30. Rao, R.V., V.J. Savsani, and D. Vakharia, Teaching–learning-based optimisation: An optimisation method for continuous non-linear large scale problems. *Information Sciences,* 2012. **183**(1): pp. 1–15.

31. Faramarzi, A., et al., Equilibrium optimiser: A novel optimisation algorithm. *Knowledge-Based Systems*, 2019: 105190.

32. Mirjalili, S., S.M. Mirjalili, and A. Lewis, Grey wolf optimizer. *Advances in Engineering Software,* 2014. **69**: pp. 46–61.

33. Li, S., et al., Slime mould algorithm: A new method for stochastic optimisation. *Future Generation Computer Systems*, 2020.

34. Faramarzi, A., et al., Marine predators algorithm: a nature-inspired metaheuristic. *Expert Systems with Applications,* 2020: p. 113377.

2 Metaheuristic Algorithms for Healthcare

Open Issues and Challenges

2.1 RESEARCH ISSUES IN HEALTHCARE

The healthcare system is currently a topic of significant investigation to make life easier for those who are disabled, old, or sick, as well as for young children [1]. The emphasis of the healthcare system has evolved throughout time due to emerging several beneficial technologies, such as personal digital assistants (PDAs), data mining, the internet of things, metaheuristics, fog computing, and cloud computing [2]. There are new innovative sensors that can collect more detailed data, but concluding the situation based on the collected data requires a sophisticated tool capable of forecasting and recognizing. Therefore, data mining and metaheuristics can contribute to the development of an intelligent and smart healthcare system, particularly for data analytics. Because metaheuristics can be used to solve problems associated with data mining or enhance the efficiency of data mining algorithms for healthcare, they can typically be used to assist us in analysing unknown as well as known data [1]. In most cases, metaheuristics [1] can be considered part of machine learning and soft computing. This chapter discusses, at the outset, the research issues of healthcare. Then, the role of metaheuristic algorithms for some healthcare optimization problems will be discussed. Last but not least, some challenges and future directions will be described in the next section. Finally, a summary and future trends are presented.

The development of computer and network technologies has provided us with several options for constructing a practical information system that can be utilized in our daily lives. In recent decades, significant developments have been made to not only other information systems but also healthcare systems. As a result of the widespread availability of modern computing and networking technology, various applications of healthcare that were previously unable to be realized may now become commonplace items. For example, in [3], a new system, known as a modern medication dispensing system, has been proposed for dispensing drugs in the healthcare system. Modern technologies like fog computing, cloud computing, deep learning, machine learning, and metaheuristics can be used to develop more accurate and complete Healthcare systems. Generally, healthcare includes several research issues which still open in front of the researchers for reaching better solutions. Some

DOI: 10.1201/9781003325574-2

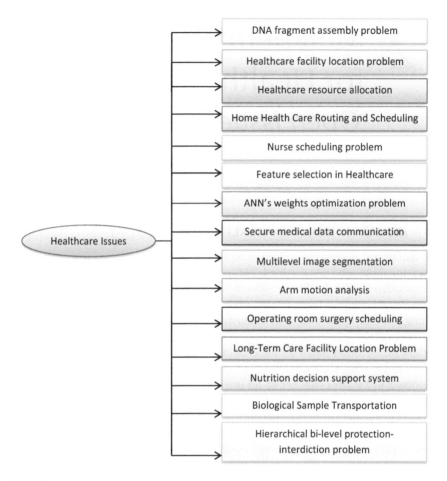

FIGURE 2.1 Depiction of some research issues in healthcare.

of these issues are DNA fragment assembly problems, medical image segmentation problems, medical feature selection problems, scheduling problems, classification problems to detect several diseases like breast cancer and heart diseases, and others depicted in Figure 2.1. Several metaheuristic algorithms have observed all those problems, but still, more research is needed to achieve better outcomes. In the rest of the chapter, we will review the roles of metaheuristics in tackling some of those Healthcare issues.

2.2 METAHEURISTICS FOR HEALTHCARE

Several recent studies on the healthcare system have used metaheuristics to solve data-mining problems, such as clustering for unknown data, classification for some unknown data, a rule to look for interesting patterns in the data, in addition to several other problems. A few of these studies will be reviewed in the next sections.

2.2.1 WHALE OPTIMIZATION ALGORITHM

Over the last five years, the whale optimization algorithm (WOA) has been applied to several optimization problems in the healthcare system. It could achieve strong outcomes compared to several traditional and modern optimization techniques. In [4], the WOA based on the overlap-layout-consensus (OLC) method has been adapted for solving the DNA fragment assembly problem (DFAP). WOA was integrated with a local search method to improve its exploitation operator when tackling the DFAP. The empirical outcomes have shown the superiority of WOA in comparison to several heuristics and metaheuristic algorithms.

The focus of [5] was on adapting the whale optimization problem for the appointment scheduling problem in the hospital under the fairness approach. In [6], the authors proposed a method known as Jaya–whale optimization (JWO), which integrates the Jaya algorithm with the whale optimization algorithm (WOA) and makes use of homomorphic encryption to initiate secure data transmission for healthcare data in the cloud. The original data are kept intact by utilizing the JWO algorithm that was proposed in order to generate a data protection (DP) coefficient.

In [7], the authors present an innovative method based on WOA and naive Bayes (NB) classifier for processing healthcare data that anticipate important information while using only a minimal amount of computational resources. The primary goal is to decrease the time required for processing while simultaneously improving accuracy and expanding the range of data types that can be accepted. This method makes use of a hybrid algorithm that employs the WOA as a feature selection technique in the first phase of the project, which attempts to reduce the total number of features used for big data. Following that, the second phase involves the application of the NB classifier to the process of real-time data classification.

In [7], it has been revealed that an innovative technique for engine health monitoring can be accomplished by combining the WOA with an artificial neural network. The WOA has trained the multilayer perceptron (MLP) neural network instead of the backpropagation algorithm to improve convergence speed and avoid getting stuck into local minima when detecting breast cancer [8]. Furthermore, the WOA has been integrated with the slime mould algorithm to overcome the image segmentation problem for COVID-19 chest X-ray images under Kapur's entropy as an objective function. Figure 2.2 depicts some optimization problems in healthcare solved recently by WOA.

2.2.2 GREY WOLF OPTIMIZER (GWO)

The home healthcare routing and scheduling problem (HHCRSP) is presented in [9] by considering the customers' priorities and the unpredictability of the times. In this work, a fuzzy multi-objective optimization model has been developed to maximize the total priority of all clients who are visited while simultaneously minimizing the cost of providing service. Jimenez's method was applied in order to modify the fuzzy parameters that are contained within the model. To solve the model, the authors suggested using a discrete multi-objective grey wolf optimizer, namely DMOGWO. The heuristic rule of priority maximum-earliest service was intended to be used

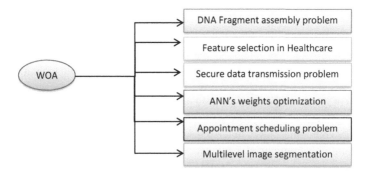

FIGURE 2.2 Some solved healthcare optimization problems using WOA.

in the DMOGWO to provide the initial solution. In order to adapt the continuous GWO algorithm to work in discrete search space, an individual position updating mechanism was built. After that, three different neighbourhood structures are built to improve the GWO algorithm's ability to do local searches.

In addition, GWO was applied to deal with the population districting problem in the health system [10]. In [11], the GWO algorithm has been combined with the NB classifier for diagnosing heart disease. GWO is employed to automatically identify the weights of the attributes of the NB classifier. These weights are then put to use in order to maximize NB's classification accuracy. Also, GWO could be employed to find the optimal subset of features that maximizes the performance of the machine learning algorithms when diagnosing diseases in healthcare. In [12], GWO was enhanced by the genetic algorithm to generate diversified initial positions, and then these positions are updated by GWO to get the optimal subset of features for getting better classification accuracy when tackling disease diagnosis problems. GWO could be employed as a learning optimizer for adjusting the weights of the artificial neural network (ANN) to accelerate the convergence speed by avoiding getting stuck into local minima, which the backpropagation algorithm might cause when predicting some diseases in the healthcare system [13, 14]. The multilevel medical image segmentation problem is another problem that could be tackled by GWO [15]. Figure 2.3 depicts some optimization problems recently solved by GWO in Healthcare.

2.2.3 Genetic Algorithms

Genetic algorithms (GAs) are among healthcare's most often used algorithms. Due to the fact that the solutions of a genetic algorithm for optimization can be expressed as integers or binary numbers, several studies [1] have employed GA to solve scheduling challenges in healthcare, such as minimizing patients' waiting time. The focus of [16] is on applying genetic algorithms to the scheduling problem of nurses to find a better schedule that improves the emergency department flow, allowing for significant reductions in patient queueing time compared to the scheduling plan created by hand. To improve patient care, a subsequent study [17] addressed multiple objectives

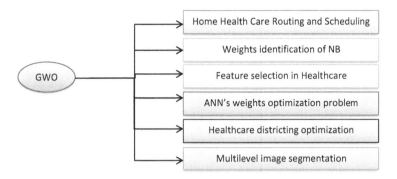

FIGURE 2.3 Some solved healthcare optimization problems using GWO.

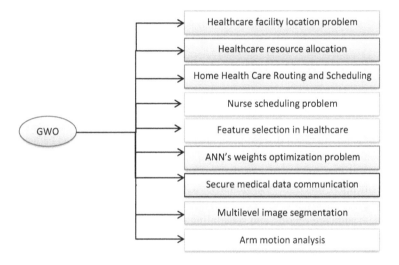

FIGURE 2.4 Some solved healthcare optimization problems using GA.

simultaneously (e.g., total patient waiting time and doctor scheduling) to optimize the effect of medical resources and reduce wasted expenses. A strategy that has shown promise in using metaheuristics for classification problems in healthcare or smart home systems is using GAs to identify better weights for adjusting classifiers or training classification algorithms. The authors of [18] employed GAs to identify improved weights for classifying human behaviours. In addition, using GAs, it is possible to improve the performance of a classification algorithm by selecting suitable features from the raw data (i.e., by reducing the number of dimensions) to accelerate the speed with which the classification algorithm operates [1]. There are several other applications for GA in healthcare, such as arm motion analysis [19], healthcare resource allocation problem [20], multilevel medical image segmentation problem [21], and healthcare facility location problem [22]. Figure 2.4 lists some healthcare optimization problems solved by GA.

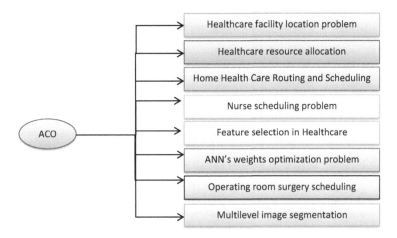

FIGURE 2.5   Some solved healthcare optimization problems using ACO.

2.2.4   ANT-COLONY OPTIMIZATION ALGORITHM

The studies [1] indicated that ant colony optimization can improve the results of a classification algorithm when applied to healthcare classification problems. ACO was used in conjunction with a fuzzy rule to classify the elements of hepatitis [1]. ACO was integrated with linear discriminant analysis (LDA) to comprehend better stomach cancer data obtained through endoscopy [1]. More specifically, in this hybrid technique, ACO-LDA, LDA performs the function of clustering the data while ACO does the classification function. Additionally, ACO has been adapted for tackling scheduling problems in healthcare. For example, in [23], ACO was adapted for tackling the home healthcare routing and scheduling problem to optimize the planning of the caregivers which was manually performed to optimize the planning of caregivers in healthcare.

Also, ACO could be employed as a trainer to the feed-forward neural network for finding the near-optimal weights which will diagnose several diseases in healthcare. For example, the ACO was employed for training the neural network for the classification problem of three diseases: heart, cancer, and diabetes [24]. More than that, ACO was applied as a feature selection technique to find the near-optimal subset of features, which will maximize the performance of the deep/machine learning algorithms, as discussed in several studies in the literature [25, 26]. There are several additional healthcare applications for ACO in healthcare scheduling [27]. Figure 2.5 lists some healthcare optimization problems solved by ACO.

2.2.5   PARTICLE SWARM OPTIMIZATION

Particle swarm optimization (PSO) is utilized to determine the weight of features for a classification algorithm or for selecting the beneficial features from the data to improve the accuracy rate of this classification algorithm [1]. Also, PSO was employed in the study [28] to analyse the data of single nucleotide polymorphisms

(SNPs) of genes involved in the renin-angiotensin system to better understand the relationship between SNPs and hypertension, such as non-hypertension and hypertension. Also, CCFS is a confidence-based and cost-effective method for selecting features based on binary PSO [29]. In this study, CCFS first increases search efficiency by inventing a new updating mechanism that explicitly considers the confidence of each feature, as well as the correlation between features and categories and the historically determined frequency of each feature. Second, the classification precision, the feature reduction ratio, and the feature cost are incorporated extensively into the design of the fitness function. The application of PSO as a classification algorithm to identify breast cancer is yet another promising direction that research is moving in [1]. The researchers [1] employed statistical methods to pick useful features of the population, and then they used discrete PSO to classify the population into two groups: those who have breast cancer and individuals who do not have breast cancer.

According to the research presented in [30], PSO can also be utilized to choose the features that are acceptable for usage with other classification algorithms. In this research, PSO was used in conjunction with a neural network in this investigation to pick the features for a support vector machine (SVM), which was then used to evaluate ultrasound images of lymph nodes. In addition to its usage as a classification algorithm, PSO has been demonstrated in several studies to have the potential for use in a variety of applications within the context of the healthcare system. PSO was employed in the research described in reference [1] to train the weight of a neural network for a touch floor system that could track the position of a user. PSO was utilized in the research stated in [1] to determine the optimal position of mass rapid transit (MRT) stations for screening passengers to stop the spread of infectious diseases. This investigation's primary focus is on modelling the problem so that it can be solved by maximizing the number of people who can be screened at each key node while also striking a balance between the number of people at each development guide plan (DGP) zone. This is done to guarantee the high quality of the collected samples.

PSO could be used to figure out a better scheduling plan for the hospital, PSO was used for finding a good sequence for the operations of all the patients by treating it as a scheduling problem. This was done to improve the quality of care provided to all of the patients. PSO was able to find a superior scheduling plan to standard scheduling technologies, such as first-come, first-served. PSO was used with an evolutionary algorithm (EA) to train the weights of a recurrent neural network for a problem involving the prediction of water quantity. PSO and EA are alternately executed at each iteration of the convergence process in this hybrid algorithm. This suggests that PSO can be employed in conjunction with other metaheuristics as the data analytics of a healthcare system. Figure 2.6 lists some healthcare optimization problems solved by PSO.

2.2.6 Other Applied Metaheuristics in Healthcare

Several other metaheuristic algorithms have been applied for feature selection problems in healthcare data. Some of those algorithms are the binary sine cosine

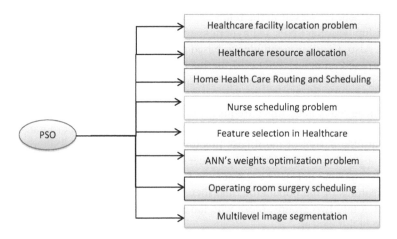

FIGURE 2.6   Some solved healthcare optimization problems using PSO.

algorithm, Harris hawks optimization algorithm, slime mould algorithm, equilibrium optimizer, and others [31, 32]. Also, the metaheuristics mentioned before could be employed to train the neural network for better classification accuracy when dealing with medical data in the healthcare system. The marine predator's algorithm (MPA) has been employed for several healthcare optimization problems. For example, in [4], the MPA has been used to train the artificial neural network for the classification problem of medical data to diagnose several diseases. The experimental findings appeared the efficiency of MPA compared to several other algorithms like particle swarm optimization (PSO) and the classical Levenberg–Marquardt (LM). In addition, MPA was applied to tackle the image segmentation problem to help the machine learning algorithms and deep learning algorithms accurately detect the persons infected with the COVID-19 virus [33]. There are several other metaheuristic algorithms for training the neural network for finding the near-optimal weights, which will fulfil better classification accuracy than previously described ones. Some of those algorithms are the bat algorithm, moth-flame optimizer, and biogeography-based optimization.

2.3   CHAPTER SUMMARY AND FUTURE TRENDS

Metaheuristics are strong technology for tackling several optimization problems in various fields, especially healthcare systems. The primary advantage of metaheuristic algorithms is their ability to find a better solution to a healthcare problem and their ability to consume as little time as possible. In addition, metaheuristics are more flexible compared to several other optimization techniques. These algorithms are not related to a specific optimization problem but could be applied to any optimization problem by making some small adaptations to become suitable to tackle it. Because metaheuristics can produce a near-optimal solution with considerably less computing time than other algorithms, they have emerged as a promising healthcare research topic in recent years, particularly for systems requiring real-time results.

In addition, the metaheuristics could be applied to several scheduling problems in healthcare like home healthcare routing and scheduling, operating room surgery scheduling, and the nurse scheduling problem. Additional problems in healthcare that metaheuristics could address include the exploration of multiple permutations to find the one that may yield a better outcome, such as the DNA fragment assembly problem and others.

Metaheuristic algorithms proved their ability to tackle several healthcare optimization problems, but unfortunately, it still suffers when tackling the high-dimensional ones. For example, the performance of the metaheuristics significantly deteriorates when training the deep neural network, which has a huge number of hidden layers due to falling into local minima and lack of population diversity. This is a hot problem that still needs a solution. Furthermore, the combinatorial problems, like scheduling problems and DFAP, could be accurately tackled by the metaheuristic because the metaheuristic produces continuous solutions which are not suitable for tackling these problems, so additional methods are employed to convert the continuous solutions into discrete ones applicable to these problems. Subsequently, in this case, the performance is not based on the metaheuristics but on the additional methods employed to adapt them for tackling the combinatorial problem. Therefore, finding a way to adapt the metaheuristics for combinatorial problems effectively is challenging. Another challenge is enhancing the performance of metaheuristic algorithms when tackling healthcare optimization problems in terms of the convergence speed to minimize the computational cost when tackling large-scale healthcare optimization problems. The additional problem is a large-scale problem. Briefly, the future research trends to reach better outcomes for the healthcare optimization problems using metaheuristics are listed as follows:

i.   Proposing a new method to preserve the diversity of the population within the optimization process.
ii. Improving the metaheuristics' convergence speed to reduce the computational cost.
iii. Improving the ability of metaheuristic algorithms to avoid getting stuck in local minima.
iv. Strengthening metaheuristics for the healthcare optimization problems on a large scale like training the deep neural network with several hidden layers.
v.   Developing an effective method to improve the performance when tackling the combinatorial healthcare optimization problems

2.4   EXERCISES

a. Why have healthcare issues been a major concern for practitioners, particularly after artificial intelligence and digital transformation?
b. What is a healthcare system? Describe its components and how they are intertwined.
c. Explain five different healthcare issues related to digitalization and patient emergency.

d. How can metaheuristics play a role in healthcare issues? Provide some justifications for existing works.

e. How be optimization concepts applied to healthcare issues? Provide some examples.

REFERENCES

1. Tsai, C.-W., et al., Metaheuristic algorithms for healthcare: Open issues and challenges. *Computers & Electrical Engineering,* 2016. **53**: pp. 421–434.
2. Deen, M.J., Information and communications technologies for elderly ubiquitous healthcare in a smart home. *Personal and Ubiquitous Computing.* 2015. **19**(3): pp. 573–599.
3. Halvorson, J.L., System for dispensing drugs in health care institutions. 1989, Google Patents.
4. Abdel-Basset, M., et al., An efficient-assembler whale optimization algorithm for DNA fragment assembly problem: Analysis and validations. *IEEE Access,* 2020. **8**: pp. 222144–222167.
5. Ala, A., et al., Optimization of an appointment scheduling problem for healthcare systems based on the quality of fairness service using whale optimization algorithm and NSGA-II. *Scientific Reports,* 2021. **11**(1): pp. 1–19.
6. Sudha, I. and R. Nedunchelian, A secure data protection technique for healthcare data in the cloud using homomorphic encryption and Jaya–Whale optimization algorithm. *International Journal of Modeling, Simulation, and Scientific Computing,* 2019. **10**(06): p. 1950040.
7. Balakrishnan, N., et al., *Aero-engine health monitoring with real flight data using whale optimization algorithm based artificial neural network technique. Optical Memory and Neural Networks,* 2021. **30**(1): p. 80–96.
8. Fang, H., et al., Automatic breast cancer detection based on optimized neural network using whale optimization algorithm. *International Journal of Imaging Systems and Technology,* 2021. **31**(1): pp. 425–438.
9. Li, Y., et al., A discrete multi-objective grey wolf optimizer for the home health care routing and scheduling problem with priorities and uncertainty. *Computers & Industrial Engineering,* 2022: p. 108256.
10. Farughi, H., S. Mostafayi, and J. Arkat, Healthcare districting optimization using gray wolf optimizer and ant lion optimizer algorithms (case study: South Khorasan healthcare system in Iran). *Journal of Optimization in Industrial Engineering,* 2019. **12**(1): pp. 119–131.
11. El Bakrawy, L., and A. Lamiaa, Grey wolf optimization and naive bayes classifier incorporation for heart disease diagnosis. *Australian Journal of Basic and Applied Sciences,* 2017. **11**(7): pp. 64–70.
12. Li, Q., et al., An enhanced grey wolf optimization based feature selection wrapped kernel extreme learning machine for medical diagnosis. *Computational and Mathematical Methods in Medicine,* 2017: Article ID 9512741.
13. Kumar, N., and D. Kumar, An improved grey wolf optimization-based learning of artificial neural network for medical data classification. *Journal of Information and Communication Technology,* 2021. **20**(2): pp. 213–248.
14. Ardabili, S., et al. Coronavirus disease (COVID-19) global prediction using hybrid artificial intelligence method of ANN trained with Grey Wolf optimizer. In *2020*

IEEE 3rd International Conference and Workshop in Óbuda on Electrical and Power Engineering (CANDO-EPE). 2020. IEEE.

15. Kumar, A., S. Singh, and A. Kumar. Grey wolf optimizer and other metaheuristic optimization techniques with image processing as their applications: a review. In *IOP Conference Series: Materials Science and Engineering.* 2021. IOP Publishing.

16. Yeh, J.-Y. and W.S. Lin, Using simulation technique and genetic algorithm to improve the quality care of a hospital emergency department. *Expert Systems with Applications,* 2007. **32**(4): pp. 1073–1083.

17. Priya, G.N., P. Anandhakumar, and K. Maheswari. Dynamic scheduler: a pervasive healthcare system in smart hospitals using RFID. In *2008 International Conference on Computing, Communication and Networking.* 2008. IEEE.

18. Fatima, I., et al. Classifier ensemble optimization for human activity recognition in smart homes. In *Proceedings of the 7th International Conference on Ubiquitous Information Management and Communication.* 2013.

19. Obo, T., et al., Arm motion analysis using genetic algorithm for rehabilitation and healthcare. *Applied Soft Computing,* 2017. **52**: pp. 81–92.

20. Feng, W.-H., et al., A multiobjective stochastic genetic algorithm for the pareto-optimal prioritization scheme design of real-time healthcare resource allocation. *Operations Research for Health Care,* 2017. **15**: pp. 32–42.

21. Hilali-Jaghdam, I., et al., Quantum and classical genetic algorithms for multilevel segmentation of medical images: A comparative study. *Computer Communications,* 2020. **162**: pp. 83–93.

22. İşbilir, M., A Genetic algorithm for healthcare facility location problem. 2016, Middle East Technical University.

23. Decerle, J., et al., A hybrid memetic-ant colony optimization algorithm for the home health care problem with time window, synchronization and working time balancing. *Swarm and Evolutionary Computation,* 2019. **46**: pp. 171–183.

24. Socha, K., and C. Blum, An ant colony optimization algorithm for continuous optimization: Application to feed-forward neural network training. *Neural Computing and Applications,* 2007. **16**(3): pp. 235–247.

25. Sweetlin, J.D., et al., Feature selection using ant colony optimization with tandem-run recruitment to diagnose bronchitis from CT scan images. *Computer Methods and Programs in Biomedicine,* 2017. **145**: pp. 115–125.

26. Fahad, L.G., et al., Ant colony optimization-based streaming feature selection: an application to the medical image diagnosis. *Scientific Programming,* 2020: Article ID 1064934.

27. Behmanesh, R., et al., Advanced ant colony optimization in healthcare scheduling. *Evolutionary Computation in Scheduling,* 2020: pp. 37–72.

28. Wu, S.-J., et al., particle swarm optimization algorithm for analyzing SNP–SNP interaction of renin-angiotensin system genes against hypertension. *Molecular Biology Reports,* 2013. **40**(7): pp. 4227–4233.

29. Chen, Y., et al. An effective feature selection scheme for healthcare data classification using binary particle swarm optimization. In *2018 9th International Conference on Information Technology in Medicine and Education (ITME).* 2018. IEEE.

30. Chang, C.-Y., C.-T. Lai, and S.-J. Chen. Applying the particle swarm optimization and Boltzmann function for feature selection and classification of lymph node in ultrasound images. In *2008 Eighth International Conference on Intelligent Systems Design and Applications.* 2008. IEEE.

31. Too, J. and S.J. Mirjalili, General learning equilibrium optimizer: A new feature selection method for biological data classification. *Applied Artificial Intelligence,* 2021. **35**(3): pp. 247–263.

32. Elgamal, Z.M., et al., An improved Harris hawks optimization algorithm with simulated annealing for feature selection in the medical field. *IEEE Access,* 2020. **8**: p. 186638–186652.

33. Abdel-Basset, M., et al., A hybrid COVID-19 detection model using an improved marine predators algorithm and a ranking-based diversity reduction strategy. *IEEE Access,* 2020. **8**: p. 79521–79540.

3 Metaheuristic-Based Augmented Multilayer Perceptrons for Cancer and Heart Disease Predictions

3.1 ARTIFICIAL INTELLIGENCE AND APPLICATION IN HEALTHCARE

Numerous studies have been conducted to predict, detect and classify diseases like heart attacks, cancer, and several others. Heart attacks, also known as cardiovascular disease (CVD), according to the World Health Organization (WHO), are one of the leading causes of death when the heart is unable to pump oxygenated blood throughout the body [1]. On the other hand, cancer occurs when cells in the body become abnormal and begin to grow out of control. Cells are the smallest building blocks that make up a human being's body. Normal cells reproduce when the body requires them and die when the body does not require them any longer (e.g., cancer cells). Cancer is made up of abnormal cells that grow and multiply even though the body does not want them to. In the majority of cancers, the aberrant cells multiply and grow into a lump or mass known as a tumour. If cancer cells remain in the body for an extended period, they can expand into (invade) surrounding areas. They have the potential to spread to other places of the body (metastasis) [2]. Recent research has widely employed machine learning methods, especially artificial neural networks, to prevent any such diseases, mostly by identifying anomalies in the human cell to pre-treat [2, 3].

Artificial neural networks (ANNs) are one of the most significant breakthroughs in the field of computational intelligence. They mostly handle classification problems by simulating the neurons of the human brain. In the literature, various varieties of ANNs have been proposed, including a feed-forward network [4], recurrent neural network [5], radial basis function (RBF) network [6], Kohonen self-organizing network [7], and spiking neural networks [8]. Regardless of their distinctions, ANNs are common in learning. The ability of a neural network to learn from experience is referred to as learning. Similar to real neurons, ANNs are equipped with mechanisms for self-adapting a given set of inputs. There are two types of learning that are frequently used in this environment: supervised and unsupervised [9]. In supervised, feedback from an external source is delivered to the ANN. However, in unsupervised learning,

DOI: 10.1201/9781003325574-3

an ANN adjusts to its inputs (learns) without the benefit of additional external feed-back. In the literature, there are two distinct categories of learning methods: deter-ministic and stochastic. Gradient-based and back-propagation (BP) algorithms are deterministic [10]. On the other hand, stochastic trainers employ stochastic optimiza-tion algorithms to maximize an ANN's performance.

The advantages of deterministic algorithms are their simplicity and rapidity of response. These algorithms, on the other hand, begin with a solution and steer it toward an optimal solution. Although the convergence of these algorithms occurs relatively quickly, the quality of the resulting solution is strongly dependent on the quality of the initial solution. Furthermore, there is a high possibility of trapping in local optima. Local optima are sub-optimal solutions in a search space that are some-times incorrectly thought to be the global optimum. The difficulty, in this case, is the unknown number of times a trainer must be restarted with multiple initial solutions to determine the global optimum [9]. On the other hand, stochastic trainers begin the training process with random solutions and evolve them throughout the opti-mization process. A major advantage of such methods is that they are quite effective at avoiding local optima [9]. However, they are typically significantly slower than deterministic alternatives. Because of their high local optima avoidance, stochastic training approaches have recently received a lot of attention, according to the litera-ture. Thus, this chapter demonstrates the influence of training the feed-forward neural network using six well-established metaheuristic algorithms to predict early cancer and heart diseases.

This chapter discusses, at the outset, the feed-forward neural network (FNNs) and multilayer perceptron (MLP). Then, the adaptation of metaheuristic algorithms to train this neural network is discussed. Lastly, the experimental findings, based on various performance metrics, are presented to illustrate the effectiveness of six investigated metaheuristics. Finally, a chapter summary is presented.

3.2 FEED-FORWARD NEURAL NETWORK

FNNs, as previously described, are ANNs with just one-directional connections between their neurons. The neurons in this sort of ANNs are organized in various par-allel layers. The first layer is always referred to as the input layer, while the last layer is referred to as the output layer. The hidden layers refer to any additional layers that exist between the input and output layers. As seen in Figure 3.1, an FNN with one hidden layer is referred to as an MLP.

The output of MLP is computed according to the following steps based on the provided inputs, weights, and biases [11]. First, the input of each hidden neuron, referred to as the weighted sums of inputs, is computed according to the following formula:

$$s_j = \left(\sum_{i=0}^{i=n} W_{ij} \cdot X_i \right) - \theta_j \tag{3.1}$$

where n denotes the number of input neurons, W_{ij} indicates the connection weight between the ith neuron in the input layer and the jth neuron in the hidden layer, θ_j

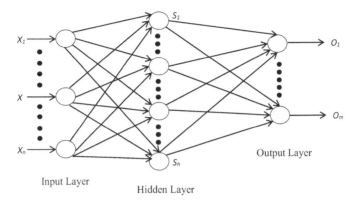

FIGURE 3.1 Multilayer perceptron with one hidden layer.

indicates the bias of the jth hidden neuron, and X_i indicates the ith input. After that, the output of each hidden neuron, or the inputs to the output layer, is computed according to the following formula:

$$S_j = \frac{1}{1+e^{-s_j}}, j = 1\ldots\ldots h \tag{3.2}$$

where h denotes the number of hidden neurons. Finally, the final outputs are computed as follows in relation to the estimated outputs of the hidden neurons:

$$O_k = \left(\sum_{j=1}^{j=h} W_{jk} \cdot S_j \right) - \theta'_k, \ k = 1,2,3,\ldots\ldots,m \tag{3.3}$$

$$O_k = \frac{1}{1+e^{-o_k}}, \tag{3.4}$$

where W_{jk} indicates the connection weight between the jth neuron in the hidden layer and the kth neuron in the output layer, θ_k indicates the bias of the kth output neuron.

As illustrated in (3.1) through (3.4), weights and biases define the ultimate output of MLPs given their inputs. The precise definition of training MLPs is finding appropriate weights and biases in order to produce a desirable relationship between the inputs and outputs. Thus, in this chapter, the performance of six well-known stochastic metaheuristic algorithms, which have strong capabilities to avoid getting stuck into local minima in comparison to the deterministic algorithms, will be investigated to show whether any of them is able to reach the weights and biases which will help in predicting cancer and heart attacks.

3.2.1 THE PROBLEM REPRESENTATION

The issue representation is the first and most critical stage in training an MLP using metaheuristics. In other words, the challenge of training MLPs should be framed in a metaheuristics-friendly manner [12]. As stated previously, the weights and biases are the most critical variables in training an MLP. A trainer should find a set of weights and biases that provides the highest accuracy in classification/approximation/prediction. As such, the variables, in this case, are the weights and biases. Due to the fact that the stochastic methods accept variables in the form of a vector, the variables of an MLP are supplied in the following manner [9]:

$$\vec{V} = \left[\vec{W}, \vec{\theta}\right] = \left[W_{1,1}, W_{1,2}, W_{1,3}, \ldots, W_{n,n}, \theta_1, \theta_2, \theta_3, \ldots, \theta_h\right] \tag{3.5}$$

After specifying the variables, we must define the respective metaheuristic's objective function. As indicated previously, training an MLP aims to achieve the maximum possible accuracy in classification, prediction, or approximation for both training and testing samples. The mean square error, abbreviated as MSE, is a frequently used measure for evaluating MLPs. This metric applies a given set of training samples to the MLP and calculates the difference between the desired output and the value received by the MLP:

$$MSE = \sum_{i=1}^{i=m} \left(o_i^D - d_i^D\right)^2 \tag{3.6}$$

where d_i^D refers to the desired output of each ith input row when the Dth training sample is employed, and o_i^D is the actual output of each ith input row when the Dth training sample is employed. MLPs must be able to classify, predict, or approximate the whole training samples to be effective. The performance of MLPs is therefore evaluated using the average of MSE for all training samples, as shown in the following formula.

$$\overline{MSE} = \frac{\left(\sum_{k=1}^{s}\sum_{i=1}^{i=m}\left(o_i^D - d_i^D\right)^2\right)}{s} \tag{3.7}$$

where s represents the training sample number and m represents the number of outputs.

In this chapter, the performance of six well-known stochastic algorithms like MPA, GWO, WOA, EO, DE, and TLBO will be employed as a trainer to the MLP for finding the weights and biases, which will minimize the average MSE of all heart attack and cancer samples. Training an MLP with the stochastic algorithm mentioned before is depicted in Figure 3.2 as an overall process. From this figure, it appears that the stochastic algorithm sends weights and biases to the MLP algorithm and receives an average MSE for all of the training samples. To minimize the average MSE of all training samples (heart attack and cancer datasets), the stochastic methods iteratively adjust weights and biases, even reaching the termination condition.

In addition, in this section, we will illustrate the pseudo code of WOA after adaptation to tackle the problem of training the MLP (see Algorithm 3.1). The other

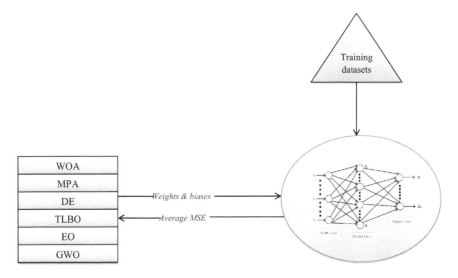

FIGURE 3.2 An optimization algorithm sends weights and biases to an MLP and takes average MSE as an objective value.

metaheuristic algorithms will be adapted using the same pattern. The whale optimization algorithm has been developed for tackling the continuous optimization problem, and could successfully tackle several optimization problems. Therefore, in this chapter, we will investigate its performance when training the MLP for detecting cancer and heart diseases. For more detail about the whale optimization algorithm, see Chapter 1. The optimization process of the WOA tries to update the initialized solutions for reaching better one. Therefore, before starting the optimization process, a number of N solutions with d dimensions, representing weights and biases, will be generated randomly at the interval of 10 and −10 as upper and lower bounds described earlier as the first step in the pseudocode of WOA adapted for training the MLP. After that, the current iteration will be set to 1, and the optimization process will be repeatedly continued even satisfying the termination condition, which is based on satisfying the maximum iteration. Finally, after satisfying the termination condition, the best-so-far solution $\overrightarrow{X^*}$ will be returned to construct the final trained MLP model, namely WOA_DNN for detecting the cancer and heart diseases. Figure 3.3 presents the flowchart of adapting any metaheuristic for training the parameters of MLP.

Algorithm 3.1 The pseudocode of WOA

Output: $\overrightarrow{X^*}$
1. Initialize N individuals
2. Find the best whale $X*$
3. $t = 1$
4. **while** $(t < T)$
5.   **for** each i whale

6.   Update a, A, p, C, and l
7.   **if** ($p < 0.5$)
8.   **if** ($|A| < 1$)
9.   Update $\overrightarrow{X}_i(t+1)$ using Eq. (1.1)
10.   **else**
11.   Update $\overrightarrow{X}_i(t+1)$ using Eq. (1.8)
12.   **end if**
13.   **else**
14.   Update $\overrightarrow{X}_i(t+1)$ using Eq. (1.6)
15.   **end if**
16.   Check the search boundary of $\overrightarrow{X}_i(t+1)$
17.   Compute the objective value of $\overrightarrow{X}_i(t+1)$
18.   Replacing the best whale $\overrightarrow{X^*}$ with $\overrightarrow{X}_i(t+1)$.
19.   if the loss value obtained by this solution is better.
20.   **end for**
21. $t++$
22. **end while**

3.2.2 CONFUSION MATRIX

In machine learning, a confusion matrix [13] is a notion that contains information about both actual and expected classifications made by the MLP trained using a

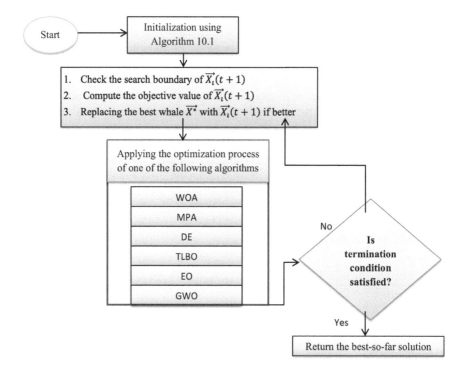

FIGURE 3.3 Flowchart of adapting metaheuristics for training the parameters of MLP.

TABLE 3.1
Confusion Matrix for Binary Classification

		Predicted class	
		Positive	**Negative**
Actual class	Positive	True positive(TP)	False negative (FP)
	Negative	False-positive (FN)	True negative (TN)

stochastic algorithm from the mentioned ones. There are two dimensions to a confusion matrix: one dimension is indexed by the actual class of an object, and the other dimension is indexed by a prediction made by the classifier about the class of an object. A confusion matrix can be defined as a summarized table showing the number of correct and incorrect predictions given by a classifier for the classification tasks. The confusion matrix is represented in the form of a square matrix, in which the column represents the actual values and the row represents the model's predicted values, as shown in Table 3.1. In this table, True positive reflects the number of positive sample unit that were correctly identified and categorized; True negative indicates the number of negative cases that were correctly identified; False positive is the number of actual negative samples classified as positive; False negative is the number of actual positive examples classified as negative. Based on this confusion matrix, some performance metrics will be extracted to evaluate the performance of the MLP trained using a stochastic algorithm; these metrics are recall or sensitivity, precision, Specificity, accuracy, and F-score and are described in detail within Table 3.2.

3.3 RESULTS AND DISCUSSION

This section will provide an extensive comparison among six metaheuristic algorithms employed to training the MLP for finding the weights and biases for three standard classification datasets obtained from the University of California at Irvine (UCI) Machine Learning Repository [14]: breast cancer, single proton emission computed tomography (SPECT) heart and SPECTF heart. These weights and biases have to optimize five performance metrics: precision, recall, accuracy, specificity, and F-score, in addition to the objective function, namely average MSE, along with the standard deviation (SD). Table 3.3 reports the properties of three standard classification datasets. Due to the stochastic nature of the metaheuristics, each one of them will be executed ten independent times, and the average of various investigated performance metrics and the average of average MSE besides SD for these average MSEs will be reported. Because this problem is a continuous optimization problem and similar to the problems tackled by six investigated metaheuristics, the parameters of these metaheuristics are set as found in the original research.

Normalization is a critical stage in the MLP process when solving datasets with properties that span many ranges. The normalizing procedure employed in this work is known as min–max normalization, described in detail in [9]. Additionally, the

TABLE 3.2
Performance Metrics Based on the Confusion Matrix

Metrics	Description	Formula
Precision	In this situation, precision tells us what percentage of the accurately anticipated cases actually turned out to be true. Precision is defined as the ratio of true positives to the sum of all true positives and false positives (true positives plus false positives). Precision examines the data to determine how many false positives were included in the mix. If there are no false positives, then the model's precision is 100%. The greater the number of FPs that are introduced into the mix, the more ugly that precision will appear.	$Precision = \dfrac{TP}{TP + FP}$
Recall	The recall rate is calculated as the ratio of the number of Positive samples that were correctly categorized as positive to the total number of positive samples. It examines the model's ability to recognize positive samples by measuring the recall. The greater the recall, the greater the number of positive samples found	$Recall = \dfrac{TP}{TP + FN}$
Accuracy	The accuracy of a machine learning classification algorithm determines how often the algorithm correctly classifies a data point. The number of correctly predicted data points out of all the data points is referred to as the accuracy of the prediction. In more formal terms, it is defined as the sum of the number of true positives and true negatives divided by the sum of the number of true positives, true negatives, false positives, and false negatives	$Accuracy = \dfrac{TP + TN}{TP + FP + TN + FN}$
Specificity	It is possible to define specificity as the ability of an algorithm or a model to predict a true negative for each available category. Or, simply, specificity, as opposed to recall, can be defined as the number of negatives returned by our machine learning model (ML model). In literature, it is referred to as the true negative rate	$Specificity = \dfrac{TN}{TN + FP}$
F-score	The harmonic mean of recall and precision will be provided by this score. The F-score is calculated mathematically as the weighted average of the recall and precision scores on a test. The best possible value for F is 1, while the poorest possible value is 0.	$F - score = \dfrac{Precision \cdot Recall}{Precision + Recal}$
g-mean	When it comes to binary classification, g-mean is defined as the squared root of the product of the Sensitivity and Specificity of the classification.	$g - mean = \sqrt{\dfrac{TP}{P} \times \dfrac{TN}{N}}$ where P indicates the number of actual true positives. And N indicates the number of actual true negatives.

TABLE 3.3
Classification Datasets

Classification datasets	Number of attributes	Number of samples	Number of training samples	Number of testing samples	Number of classes
Breast cancer	31	569	455	114	2
SPECT	23	267	214	53	2
SPECTF	45	267	214	53	2

TABLE 3.4
MLP Structure for Each Dataset

Classification datasets	Number of attributes	Number of input neurons	Number of hidden neurons	Number of output neurons	MLP structure
Breast cancer	31	30	61	1	30-61-1
SPECT	23	22	45	1	22-45-1
SPECTF	45	44	89	1	44-89-1

structure of MLPs plays a significant role in the experiments. Therefore, the structure of each MLP used in each dataset with computing the number of hidden nodes using $2 \times N + 1$, where N is the number of features (inputs), is presented in Table 3.4.

3.3.1 BREAST CANCER DATASET

The breast cancer dataset, a common dataset in the literature, has 30 attributes, 455 training samples, 114 test samples, and 2 classes. As can be seen in Table 3.4, an MLP with a structure of 30-61-1 is selected for the breast cancer dataset. As a result, every search agent of the trainers (metaheuristic algorithms) in this study has a total of 1953 variables to be optimized. The trainers solved this dataset ten independent times, and the statistical results are provided in Table 3.5. From this table, GWO could be better than all the other metaheuristics in terms of the average MSE and SD on the training instance, followed by TLBO and EO as the second and third best trainers, while WOA is classified as the worst one, as shown in Figure 3.4. In addition, to further show the trainers' performance, five performance metrics are used to evaluate the MLP trained using a stochastic algorithm. These metrics are accuracy, recall, specificity, precision, F-score, and g-mean. The average values of these metrics obtained after ten independent times are presented in Table 3.5, which shows that the performance of TLBO is better for all performance metrics over the training samples, followed by MPA as the second-best one. In addition, Figure 3.5 is presented to depict the convergence speed of each optimizer. This figure demonstrates that TLBO has a faster convergence rate at the beginning of the optimization process, while GWO outperforms at the end.

TABLE 3.5
Experimental Findings for Breast Cancer

	GWO		DE		WOA		MPA		TLBO		EO	
	Training	Testing	Training	Testing	Training	Testing	Training	Testing	Training	Testing	Training	Testing
Average MSE	**0.00576**		0.04492		0.13909		0.01037		0.00699		0.00812	
SD	**0.00207**		0.01029		0.04486		0.00466		0.00388		0.00325	
Accuracy	0.96751	0.93167	0.94304	0.91000	0.78397	0.76583	0.98776	**0.95333**	**0.99536**	0.94750	0.95612	0.91000
Recall	0.94710	0.83649	0.89058	0.81351	0.61232	0.51757	0.95652	0.81622	**0.97536**	0.85270	0.92536	**0.87432**
Specificity	0.96912	0.90633	0.93702	0.89104	0.76290	0.72783	0.97502	0.89483	**0.98585**	0.91369	0.95662	**0.92300**
Precision	0.96829	0.91611	0.93980	0.89809	0.76856	0.73885	0.98131	0.92282	**0.99057**	**0.92936**	0.95628	0.91477
F-score	0.95724	0.88016	0.91621	0.85756	0.68487	0.60371	0.97198	0.88140	**0.98528**	**0.89791**	0.94053	0.89068
g-mean	96.0000	89.5360	92.3733	87.3195	72.0800	67.1134	97.6266	90.1030	**98.8000**	**91.1340**	94.4800	89.6391

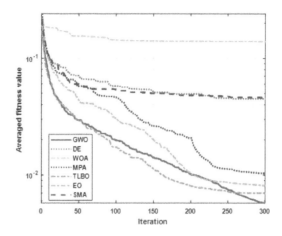

FIGURE 3.4 Convergence curve.

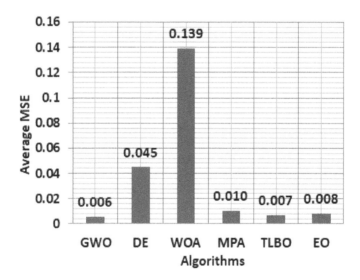

FIGURE 3.5 Depiction of average MSE.

Over the testing samples, TLBO could be better for three performance metrics: precision, F-score, and g-mean, and DE is better for recall and specificity, while MPA accurately achieves accuracy. From this analysis, over the training instance, it is concluded that GWO has strong performance for optimizing the average MSE, while TLBO is better for all performance metrics. On the other side, TLBO could be better for most performance metrics over the testing samples. As a result, TLBO could be employed to train the MLP to reach the weights and biases that will accurately detect breast cancer patients.

3.3.2   SPECT HEART DATASET

This dataset is a well-known dataset in the literature that presents a number of challenges. There are 22 attributes in this dataset, as well as 214 training examples, 53 test samples, and two classes. Specifically, as shown in Table 3.3, an MLP with a structure of 22-45-1 is selected to be trained by the metaheuristics technique described previously to solve this dataset. The consequence is that each search agent of the trainers (metaheuristic algorithms) in this study has a total of 1081 variables to optimize. In total, the trainers solved this dataset ten separate times, with the statistical findings supplied in Table 3.6 for your convenience. The average MSE and SD on the training instances affirm that GWO is superior to all other metaheuristics, according to this table and Figure 3.7, also followed by TLBO and EO as the second and third best trainers, respectively. WOA is the worst trainer, based on this table's findings.

Furthermore, Figure 3.6 illustrates the convergence speed of each optimizer. This figure reveals that TLBO has a faster convergence rate at the start of the optimization process, whereas GWO outperforms at the end. In addition, six performance metrics are utilized to evaluate the MLP trained using a stochastic method to demonstrate the trainers' overall performance. Accuracy, recall, specificity, precision, F-score, and g-mean are measured metrics. As shown in Table 3.6, the average values of these metrics acquired after ten independent runs are superior for all performance metrics compared to the training samples, with TLBO being the top performer overall. Over the testing samples, DE is better than all the others for all performance metrics. Based on this data, it can be determined that GWO has excellent performance for optimizing the average MSE over the training instance. In contrast, TLBO has superior performance for all performance measures. On the other hand, DE is superior to the testing samples in terms of all performance indicators. As a result, TLBO and DE could be used to train the MLP to get the weights and biases necessary for accurately detecting SPECT heart disease.

3.3.3   SPECTF HEART DATASET

This dataset has 44 attributes, 214 training samples, 53 test samples, and two classes. As can be seen in Table 3.3, an MLP with a structure of 44-89-1 is picked to be trained by the metaheuristics approach previously discussed to solve this dataset. As a result, each search agent of the metaheuristic algorithms in this work has 4095 variables that need to be optimized. Each algorithm will be run ten independent times on this dataset, and the statistical outcomes are presented in Table 3.7. According to this table, GWO appears to be superior to all other metaheuristics in terms of the average MSE, followed by EO, MPA, and TLBO as the second, third, and fourth best trainers, respectively, while WOA is considered to be the poorest trainer. In addition, to demonstrate the effectiveness of the trainers, five performance indicators are utilized to evaluate the MLP trained using each stochastic method. The average values of these metrics acquired after ten separate times are provided in Table 3.7, which demonstrates that the performance of TLBO is superior for three performance metrics: recall, specificity, and g-mean over the training samples, while MPA is better for the others. Over the testing samples, WOA could reach the weights, which will

TABLE 3.6
Experimental Findings for SPECT Heart Disease

	GWO		DE		WOA		MPA		TLBO		EO	
	Training	Testing	Training	Testing	Training	Testing	Training	Testing	Training	Testing	Training	Testing
Average MSE	**0.07392**		0.12607		0.12608		0.09056		0.08066		0.08505	
SD	**0.00983**		0.00422		0.02069		0.01060		0.01164		0.00950	
Accuracy	0.92384	0.85366	0.93198	**0.94309**	0.84593	0.86992	0.91221	0.85366	**0.93314**	0.79675	0.92674	0.84553
Recall	0.25610	0.10256	0.30488	**0.23077**	0.01463	0.02564	0.23659	0.02564	**0.36341**	0.17949	0.21951	0.20513
Specificity	0.83898	0.75000	0.84994	**0.79483**	0.78157	0.73773	0.83400	0.73403	**0.86124**	0.75500	0.83272	0.77112
Precision	0.87927	0.79815	0.88879	**0.86238**	0.81137	0.79832	0.87116	0.78920	**0.89541**	0.77496	0.87709	0.80555
F-score	0.48267	0.29056	0.51141	**0.46131**	0.07676	0.08782	0.45233	0.08663	**0.57010**	0.30735	0.44571	0.39912
g-mean	79.5305	67.2839	81.1267	**77.1604**	68.5915	66.6666	78.2159	65.4321	**82.3474**	64.8148	79.0610	69.1358

TABLE 3.7
Experimental Findings for the SPECTF Heart Disease

	GWO		DE		WOA		MPA		TLBO		EO	
	Training	Testing	Training	Testing	Training	Testing	Training	Testing	Training	Testing	Training	Testing
Average MSE	**0.04954**		0.16940		0.19571		0.09307		0.09473		0.07690	
SD	0.00937		0.00623		**0.00646**		0.01073		0.03035		0.01606	
Accuracy	0.86588	0.88718	0.92412	**0.93333**	**0.97824**	0.92045	0.94294	0.92308	0.92000	0.91282	0.88765	0.85128
Recall	0.37674	0.09333	0.21628	0.10667	0.01628	0.02000	0.43488	0.06667	**0.47209**	**0.13333**	0.43023	0.14667
Specificity	0.84576	0.71747	0.82421	0.73126	0.79720	**0.80308**	0.86863	0.71968	**0.87418**	0.73321	0.86027	0.72236
Precision	0.85556	0.79300	0.87055	0.81894	0.87841	**0.85421**	**0.90414**	0.80805	0.89616	0.81096	0.87366	0.77994
F-score	0.56803	0.28334	0.40904	0.23608	0.08720	0.03371	0.63595	0.18677	**0.64772**	0.25315	0.60906	**0.30270**
g-mean	76.7136	66.6666	78.1220	70.3703	78.4037	75.3703	**84.0375**	68.5185	82.9577	69.6296	79.5305	65.5555

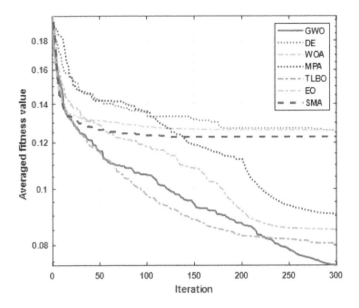

FIGURE 3.6 Convergence curve.

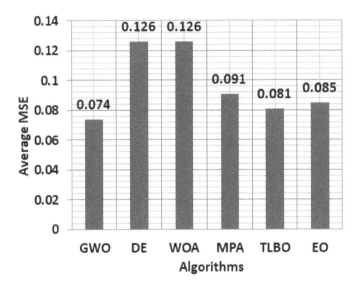

FIGURE 3.7 Depiction of average MSE.

produce better specificity, precision, and g-mean; DE could be superior for accuracy, and TLBO could be superior for recall.

Furthermore, Figure 3.8 depicts the convergence speed of each optimizer. This figure reveals that GWO has a faster convergence rate at all stages of the optimization

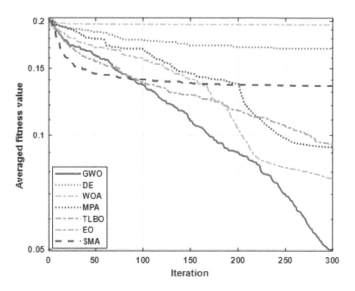

FIGURE 3.8   Convergence curve.

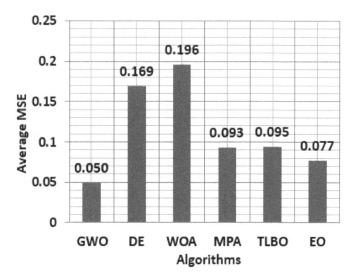

FIGURE 3.9   Depiction of average MSE.

process. In addition, the average MSE within the independent runs are computed and reported in Figure 3.9. Inspecting this figure give us a knowledge that GWO could minimize the mean square error between the measured and estimated label better than all the other optimizers. Broadly speaking, from this figure, GWO could occupy the first rank with a value of 0.05, and EO comes in the second rank with an average MSE of 0.077, while WOA is the worst one because it got the highest MSE value.

Based on these results, it can be concluded that GWO performs well for maximizing the average MSE over the training instance, whereas TLBO performs better for three performance measures: recall, specificity, and F-score. Over the testing samples, TLBO could be competitive with most of the compared metaheuristic algorithms. WOA could be better for three performance metrics: Specificity, precision, and g-mean. Therefore, TLBO could be utilized to train the MLP in order to obtain the weights and biases necessary for an accurately SPECTF heart dataset.

3.4 CHAPTER SUMMARY

This chapter demonstrates the impact of training a feed-forward neural network using six well-established metaheuristic algorithms: MPA, GWO, EO, WOA, TLBO, and EO for early cancer and heart disease prediction. Three standard classification datasets obtained from the University of California at Irvine (UCI) Machine Learning Repository: breast cancer, single proton emission computed tomography (SPECT) heart, and SPECTF heart are employed to validate the performance of MLP trained using these algorithms under various performance metrics such as precision, recall, g-mean, F-score, specificity, and accuracy. The experimental findings show that TLBO and DE are strong stochastic optimizers for training the MLP to reach the near-optimal weights and biases.

3.5 EXERCISES

a. Provide examples of successful application of different artificial intelligence concepts in the healthcare system.
b. What varieties of ANNs are prominent in the literature? How do they differ?
c. Explain a multilayer perceptron feed-forward neural network model with supporting pros and cons.
d. Name a few performance metrics commonly used to validate multilayer perceptron models. Explain them briefly
e. Explain how metaheuristics can be plugged-in into the multilayer perceptron model.
f. How can a metaheuristic model influence the final result for that neural network model?

REFERENCES

1. Terrada, O., et al. Prediction of patients with heart disease using artificial neural network and adaptive boosting techniques. In *2020 3rd International Conference on Advanced Communication Technologies and Networking (CommNet)*. 2020. IEEE.
2. Kashf, D.W.A., et al., Predicting DNA lung cancer using artificial neural network. *International Journal of Academic Pedagogical Research,* 2018. **2**: 6–13.
3. Malav, A., et al., Prediction of heart disease using k-means and artificial neural network as hybrid approach to improve accuracy. *International Journal of Engineering and Technology,* 2017. **9**(4): pp. 3081–3085.
4. Bebis, G. and M. Georgiopoulos, Feed-forward neural networks. *IEEE Potentials,* 1994. **13**(4): pp. 27–31.

5. Dorffner, G. Neural networks for time series processing. *Neural Network World*. 1996.

6. Park, J. and I.W. Sandberg, Approximation and radial-basis-function networks. *Neural Computation,* 1993. **5**(2): pp. 305–316.

7. Kohonen, T., The self-organising map. *Proceedings of the IEEE,* 1990. **78**(9): pp. 1464–1480.

8. Ghosh-Dastidar, S. and H. Adeli, Spiking neural networks. *International Journal of Neural Systems,* 2009. **19**(04): pp. 295–308.

9. Mirjalili, S., How effective is the grey wolf optimiser in training multi-layer perceptrons. *Applied Intelligence,* 2015. **43**(1): pp. 150–161.

10. Hertz, J., A. Krogh, and R.G. Palmer, *Introduction to the theory of neural computation.* 2018, CRC Press.

11. Mirjalili, S. and A.S. Sadiq. Magnetic optimisation algorithm for training multi layer perceptron. In *2011 IEEE 3rd International Conference on Communication Software and Networks.* 2011. IEEE.

12. Belew, R.K., J. McInerney, and N.N. Schraudolph. *Evolving networks: Using the genetic algorithm with connectionist learning.* 1990, Citeseer.

13. Sammut, C. and G.I. Webb, *Encyclopedia of machine learning.* 2011, Springer Science & Business Media.

14. Blake, C., *UCI repository of machine learning databases.* 1998.

4 The Role of Metaheuristics in Multilevel Thresholding Image Segmentation

4.1 MULTILEVEL THRESHOLDING IMAGE SEGMENTATION

The medical resources necessary for COVID-19 detection in many countries are still insufficient. Therefore, creating a low-cost, rapid method to successfully detect and identify the virus is important to minimize the required tests. When processing large numbers of tests, it is necessary to interpret images generated by chest X-rays precisely and promptly. This is especially true when processing a high number of tests at the same time. Symptoms of COVID-19 include bilateral pulmonary parenchymal ground-glass opacities and consolidative pulmonary opacities, which can have an irregular shape and be found in the peripheral lungs. By utilizing chest X-ray images, we hope to be able to quickly locate and remove similar tiny regions that may have the defining features of COVID-19 from them. Medical imaging techniques are crucial for illness diagnosis and patient health care. They aid in the diagnosis, treatment, and early identification of disease. Segmentation is a critical stage in medical image processing and has been widely applied in a variety of applications.

This chapter discusses Kapur's entropy fitness function employed with various metaheuristic algorithms (MHAs) for segmenting some chest X-ray images under various threshold levels. In addition, the experimental settings and chest X-ray test images are also described. Last but not least, results and discussion are also presented to illustrate the influence of MHAs on tackling this problem. Finally, a chapter summary is presented to show which investigated algorithm is preferred for segmenting chest X-ray images.

4.2 IMAGE SEGMENTATION TECHNIQUES

Several techniques have been proposed for image segmentation, including region-based, edge-based, threshold-based, and feature-based clustering [1]. Edge-based segmentation focuses on finding the boundaries of various objects in an image. This is an important stage since it assists you in identifying the characteristics of the numerous

DOI: 10.1201/9781003325574-4

objects included in the image because edges hold a great deal of information that can be used to your advantage. A common reason for the widespread use of edge detection is that it assists you in deleting undesirable and superfluous information from an image. It significantly reduces the image's size, making it easier to analyse the image. Using edge-based segmentation algorithms, edges in an image are identified based on changes in texture, contrast, grey-level (degree of detail), colour, saturation, and other attributes. Several edge-based segmentation techniques are divided into two categories: search-based edge detection and zero-crossing-based edge detection. Segmentation techniques based on regions divide the image into pieces with similar attributes. These regions consist of a collection of pixels, and the algorithm identifies them by first selecting a seed point, which may be a tiny chunk or a significant portion of the input image. Following the discovery of the seed points, a region-based segmentation algorithm would either augment them with additional pixels or decrease them in order to integrate them with other seed points. Based on that, region-based segmentation could be classified into two categories: region growing and region splitting and merging.

Clustering-based segmentation splits an image into clusters (disjoint groupings) of pixels with similar characteristics. With this technique, the pixels would be grouped together into clusters if the items inside a cluster are more similar in nature to the pixels present in other clusters. Some of the most often used clustering algorithms are fuzzy c-means (FCM), k-means, and enhanced k-means algorithms. Threshold-based segmentation is classified into bi-level thresholding (BLT) and multilevel thresholding (MLT). BLT divides an image into two classes, one comprising pixels with grey levels greater than a threshold and the other containing the remaining pixels. However, the BLT encounters a problem when an image contains more than two classes. Thus, the MLT can address this issue by incorporating the subdivision of a given image into more classes.

Traditional MLT segmentation algorithms are based on the image's grey-level histogram [12] and involve minimizing certain fitness functions such as entropy and Otsu. However, classic MLT approaches suffer from being time-consuming, particularly as the number of threshold levels increases. Additionally, they are easily immobilized at a particular place. As a result, metaheuristic algorithms (MHAs) have been widely used to improve MLT, as MLT can be regarded as an NP-hard problem. MHAs need to be adapted to minimize or maximize an objective function on account of the optimization problems. In MLT, there are three various fitness functions, including Kapur's entropy [14], fuzzy entropy [15], and Otsu function [16], which could be employed with MHAs for segmenting an image. Several MHAs have been proposed for tackling the multilevel thresholding image segmentation under various fitness functions: under Kapur's entropy, including EO [2], MPA [3], moth swarm algorithm (MSA) [4], crow search algorithm (CSA) [5], firefly algorithm (FFA) [6], and symbiotic organisms search (SOS) algorithm [7]; under fuzzy entropy, like spherical search optimizer (SSO) [8], MPA [9], and electromagnetic field optimization algorithm (EFOA) [10]; and under Otsu function, like black widow optimization (BWO) algorithm [11], grey wolf optimizer (GWO) [12], and FFA[13].

4.2.1 Objective Function: Kapur's Entropy

Kapur's entropy, which is utilized to derive the ideal threshold values by maximizing the entropy of segmented regions [14], will be discussed in detail in this section. Let's look at the mathematical model of this technique. Assuming that the threshold values that segment an image into k comparable parts are $t_0, t_1, t_2, \ldots\ldots, $ and t_k, then the Kapur's entropy seeks for maximization of the following formula till attaining the ideal threshold values:

$$T\left(t_0, t_1, t_2, \ldots\ldots, t_k\right) = T_0 + T_1 + T_2 + \ldots\ldots + T_k \tag{4.1}$$

where :

$$T_0 = -\sum_{i=0}^{t_0-1} \frac{X_i}{W_0} * ln\frac{X_i}{W_0}, \ X_i = \frac{N_i}{W}, \ W_0 = \sum_{i=0}^{t_1-1} X_I \tag{4.2}$$

$$T_1 = -\sum_{i=t_0}^{t_1-1} \frac{X_i}{W_1} * ln\frac{X_i}{W_1}, \ X_i = \frac{N_i}{W}, \ W_1 = \sum_{i=t_0}^{t_1-1} X_I \tag{4.3}$$

$$T_2 = -\sum_{i=t_1}^{t_2-1} \frac{X_i}{W_2} * ln\frac{X_i}{W_2}, \ X_i = \frac{N_i}{W}, W_2 = \sum_{i=t_1}^{t_2-1} X_I \tag{4.4}$$

$$T_k = -\sum_{i=t_k}^{L-1} \frac{X_i}{W_k} * ln\frac{X_i}{W_k}, \ X_i = \frac{N_i}{W}, W_k = \sum_{i=t_k}^{L-1} X_I \tag{4.5}$$

T_0, T_1, T_2, $\ldots\ldots$, and T_k denote the entropies of identical regions, whereas N_i is the number of pixels with an integer value equal to i. W_0, W_1, $W_2, \ldots,$ and W_k are all terms that refer to the probability of various regions in relation to the entire pixel W within an image.

4.2.2 Adaptation of Metaheuristics for Image Segmentation

Prior to initiating the optimization process, a group of N prey will be established, with each member having a number of dimensions proportional to the achieved threshold level. These dimensions will be dispersed uniformly over the search space using the following:

$$\vec{X}_i = \vec{X_L} + \left(\vec{X_U} - \vec{X_L}\right) * \vec{r} \tag{4.6}$$

Where $\vec{X_L}$, and $\vec{X_U}$ are vectors representing the upper and lower bounds of an image's grey level inside its histogram, respectively. For example, assuming that the upper bound grey level is equal to 255, the lower bound is 0, and the threshold level of 10 is required. Each solution within the population will be represented as shown

5.1	12.5	20.5	100.5	50.5	30.2	150.3	200.1	250.2	80.4

The threshold level of 10

FIGURE 4.1   Solution representation to multilevel thresholding.

5	12	20	100	50	30	150	200	250	80

Threshold level of 10

FIGURE 4.2   unordered integer threshold values.

5	12	20	30	50	80	100	150	200	250

Threshold level of 10

FIGURE 4.3   Ordered integer threshold values.

in Figure 4.1, adding the first and the last cell with the image's lower and upper grey levels. Various MHAs are used here to determine the threshold values that will separate those regions from one another. MHAs will randomly distribute their solutions within the search space, as illustrated in Figure 4.1.

After spreading the solutions within the problem's bounds, these values should be converted to integers, as each pixel in the grey image is represented by just eight bits for an integer value, and so each pixel will load only an integer value, not a decimal value. As a result, the values preceding the dot in Figure 4.2 will be used to represent the solutions to the image segmentation problem, while the numbers after the dot will be trimmed, as shown in Figure 4.3.

Following that, the integers in Figure 4.2 will be grouped in the manner shown in Figure 4.3), and Eq. (4.1) will be used to determine the quality of those threshold values in terms of Kapur's entropy..

The preceding stages distribute the problem's dimensions (number of needed threshold values) across the search space, convert them to integer values, organize them, and evaluate them using Equation (4.1). These will be repeated for each solution during the initialization process. Following that, the initialization phase is skipped, and the solution generated by a metaheuristic approach during the optimization process is simply transformed to an integer, organized, and evaluated using Equation (4.1) as shown in Figure 4.4.

Finally, the best so-far threshold values will be returned after completing the optimization process to build the segmented image. However, how will the segmented image be formed using the obtained threshold values? Assume that the original image is named A and has rows and columns count of N and M, respectively. Once the ideal threshold values are determined for any threshold level, the segmented image is formed, as shown in Algorithm 4.1.

Algorithm 4.1 segmented image generation steps (GSI)

---

1. B: is a matrix of N*M to include the pixels of the segmented image.
2. $W^* = [0 \ \bar{X}^* \ 255]$.
3. **for** i=0: N
4.   **for** j=0: M
5.   **for** m=0: t-1
6.   **if** $A(i,j) \geq W^*_{(m)}$ & &$A(i,j) \leq W^*_{(m+1)}$
7.   B(i, j)= W^*_m ;
8.   **end if**
9.   **end for**
10.   **end for**
11.   **end for**
12. **return** B;

4.2.3 PERFORMANCE EVALUATION CRITERIA

The performance measures used to monitor the algorithms' performance will be addressed briefly in this section. The standard deviation (SD), the peak signal to noise ratio (PSNR), the structured similarity index metric (SSIM), the feature similarity index (FSIM), and the fitness value under Kapur's entropy are the metrics used.

4.2.3.1 Standard Deviation (SD)

SD is used to quantify the stability of each algorithm's results over multiple runs and is determined analytically using the following equation:

$$SD = \sqrt{\frac{1}{n-1}\sum_{i=1}^{n}\left(f_i - \bar{f}\right)^2} \qquad (4.7)$$

where n indicates the number of separate runs, f_i denotes the fitness value under Kapur's entropy for the ith run, and $\bar{f}$ denotes the average of the fitness values for the independent runs. The algorithm with the lowest standard deviation is deemed to be the best.

4.2.3.2 Peak Signal-to-Noise Ratio (PSNR)

PSNR [15] is a mathematical formula that calculates the ratio of error between the original and segmented images:

$$PSNR = 10\left(\frac{255^2}{MSE}\right) \qquad (4.8)$$

MSE is a measure of the mean squared error between two images and is calculated as follows:

$$MSE = \frac{\sum_{i=1}^{M}\sum_{j=1}^{N}\left|A(i,j) - S(i,j)\right|}{M*N} \qquad (4.9)$$

where $A(i,j)$ specifies the grayscale intensity level of the segmented image, and $S(i,j)$ is the original image matrix's grey level within the row, i^{th} and column j^{th}. M and N are the rows and columns contained within the image, respectively.

4.2.3.3 Structural Similarity Index (SSIM)

PSNR ignores the image's structure because it calculates the ratio of the error between the original and segmented images. SSIM [15] is a mathematical formula that is designed to take into account the similarity, contrast distortion, and brightness of two images:

$$SSIM(O,S) = \frac{(2\mu_o\mu_s + a)(2\sigma_{os} + b)}{(\mu_o^2 + \mu_s^2 + a)(\sigma_o^2 + \sigma_s^2 + b)} \tag{4.10}$$

where μ_o denotes the average intensities of the source image and μ_s denotes the segmented image's average intensities. The standard deviations of original and segmented images are denoted by σ_o and σ_s. The covariance between the original and segmented images is denoted by σ_{os}, and a, b are two constants equal to 0.001 and 0.003, respectively. The algorithm that returns the highest SSIM value is the optimal one.

4.2.3.4 Features Similarity Index (FSIM)

FSIM [16] is a statistic for determining the similarity of features between the segmented and original images. FSIM could be expressed mathematically as follows:

$$FSIM(O,S) = \frac{\sum_{X \in \Omega} S_T(X) * PC_m(X)}{\sum_{X \in \Omega} PC_m(X)} \tag{4.11}$$

where Ω represents the entire pixel domain of an image. $S_T(X)$ defines a similarity score. The phase consistency measure is denoted by $PC_m(X)$, which is expressed as:

$$PC_m(X) = max\big(PC_1(X), PC_2(X)\big) \tag{4.12}$$

where $PC_1(X)$ and $PC_2(X)$ describe the phase consistency of two blocks, respectively:

$$S_T(X) = \big[S_{PC}(X)\big]^\alpha \cdot \big[S_G(X)\big]^\beta \tag{4.13}$$

$$S_{PC}(X) = \frac{2PC_1(X) \times PC_2(X) + T_1}{PC_1^2(X) \times PC_2^2(X) + T_1} \tag{4.14}$$

$$S_G(X) = \frac{2G_1(X) \times G_2(X) + T_2}{G_1^2(X) \times G_2^2(X) + T_2} \tag{4.15}$$

where the similarity measure of phase consistency is denoted by $S_{PC}(X)$. The gradient magnitude of the two areas $G_1(X)$ and $G_2(X)$ is denoted by $S_G(X)$. The constants, T_1, and T_2 are all present. The value of FSIM is ranged between 0 and 1, and indicates that the segmented image quality is greater when the value is higher.

4.2.4 EXPERIMENT SETTINGS

The parameter values of the investigated MHAs, the image dataset, and parameter settings are discussed in detail within this section to show the settings that the experiments were based on. Concerning the image dataset, during our experiment, we used eight COVID-19 chest X-ray images obtained from https://github.com/ieee8 023/covid-chestxray-dataset to evaluate the performance of MHAs in extracting similar regions from the medical images for illness diagnosis and patient healthcare, which are labelled as R1, R2, R3, R4, R5, R6, R7, R8, and R9. Figure 4.4 depicts the original images as well as their histograms.

After that, several experiments were conducted on this dataset to compare the performance of the seven MHAs for finding the optimal threshold values. MHAs extract the similar regions through an image over various threshold levels (K) ranging between 3 and 40. Those MHAs are WOA, DE, GWO, EO, SMA, MPA, and TLBO. In reality, the population-based metaheuristic algorithms contain two primary effective parameters that must be assigned equally to provide a fair comparison between the algorithms. Those two parameters are the maximum iteration and the population size, which are set to 50 for each test image and 25 for the overall population size. As a result, each algorithm will have only 25 solutions, each of which will be updated within 50 iterations in accordance with the updating strategy of this algorithm, and thus a fair comparison between algorithms will be achieved, with the algorithm producing the best results being declared the winner. It is worth highlighting that all algorithms are implemented in MATLAB R2019a and run on the same device with settings of 32GB RAM and an Intel CoreI i7-4700MQ CPU running at 2.40 GHz.

4.2.5 CHOICE OF PARAMETERS

Regarding the other parameters of the investigated MHAs, extensive experiments have been done in this chapter to find the more suitable parameter value for each algorithm for tackling the image segmentation problem. To do that for MPA, which has two main parameters: P and FADs, need to be estimated accurately to maximize its performance. Extensive experiments under various values for each parameter have been executed, and the average fitness values obtained under each one of those values are depicted in Figure 4.5. Inspecting this figure shows that the best values for P and FADs are of 0.01 and 0.05, respectively. EO algorithm has four main effective parameters: a_1, a_2, V, and GP, which need to be accurately estimated to maximize its performance. Therefore, various values for these parameters are investigated to show which one improves the performance of EO (see Figure 4.6). This figure concludes that the best values for these parameters, respectively, are 0.5, 5, 1, and 0.6. Regarding the parameters of DE, which are F and Cr, the best values for them are estimated by running DE under various values for each parameter. The obtained outcomes (average fitness value for each parameter) are shown in Figure 4.7. Based

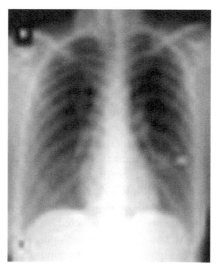

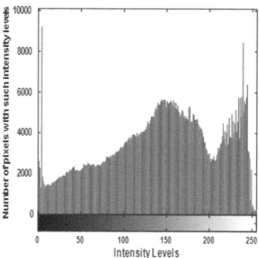

FIGURE 4.4 The original chest X-ray images and their histogram. (a) Original image, R1; (b) histogram of R1 image; (c) original image, R2; (d) histogram of R2 image; (e) original image, R3; (f) histogram of R3 image; (g) original image, R4; (h) histogram of R4 image; (i) original image, R5; (j) histogram of R5 image; (k) original image, R6; (l) histogram of R6; (m) original image, R7; (n) histogram of R7; (o) original image R8; (p) histogram of R8.

on the outcomes in this figure, we set the best values for both F and CR to 0.1 and 0.6, respectively.

Last but not least, WOA has a parameter, namely b, which is responsible for defining the logarithmic spiral shape. To estimate the optimal value for this parameter, experiments have been done under various values, including 0.5, 1, 2, 3, 4, and 5, which show that the best value for this parameter is 0.5 as elaborated in Figure 4.8.

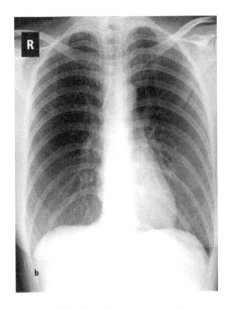

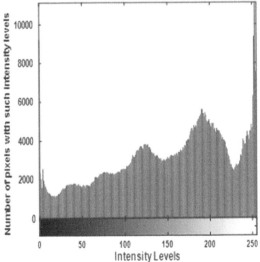

FIGURE 4.4   (Continued)

Finally, SMA has one effective parameter: z, which is estimated in Figure 4.9 under extensive experiments performed using various values for this parameter. According to this figure, the best value for this parameter is 0.05.

4.3   RESULTS AND DISCUSSION

The MHAs will be herein evaluated under various performance metrics like fitness value (F-value), PSNR, SSIM, and FSIM.

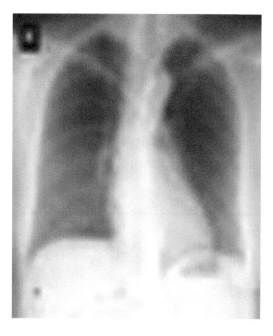

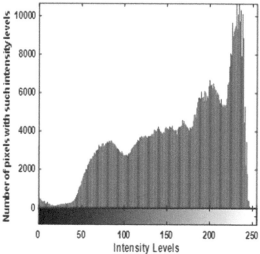

FIGURE 4.4　(Continued)

4.3.1 STABILITY

Starting with stability, to quantify the convergence between the outputs produced by each metaheuristic algorithm, the average of SD was determined on all medical test images during 30 independent runs for each one and displayed in Figure 4.10. This figure illustrates that TLBO could give more convergent outputs in comparison with the other algorithm, and SMA is the lowest stable one.

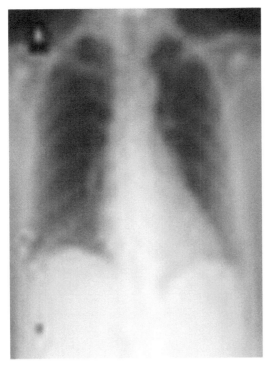

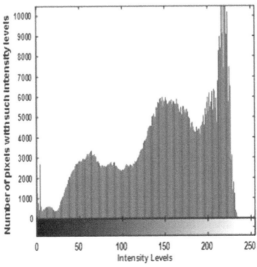

FIGURE 4.4 (Continued)

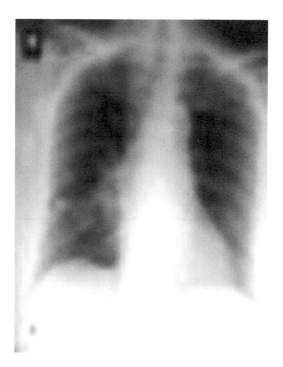

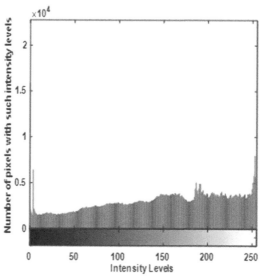

FIGURE 4.4   (Continued)

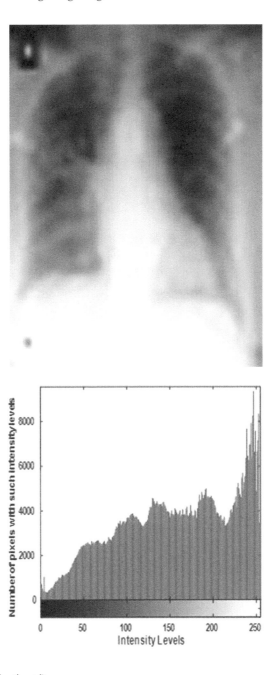

FIGURE 4.4   (Continued)

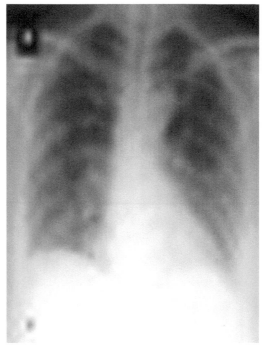

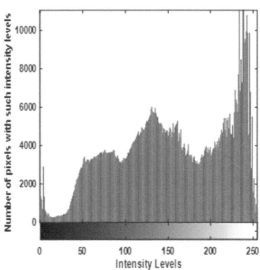

FIGURE 4.4   (Continued)

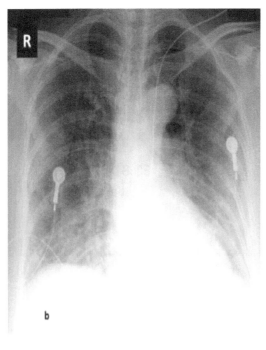

b

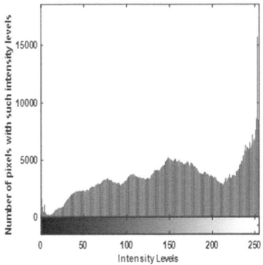

FIGURE 4.4   (Continued)

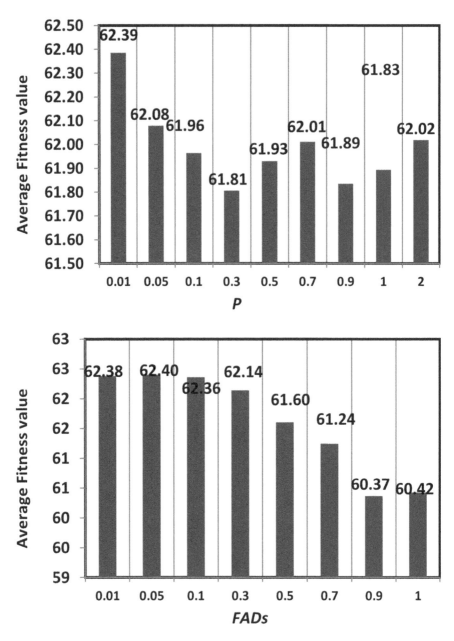

FIGURE 4.5 Tuning the parameters of MPA over R1.

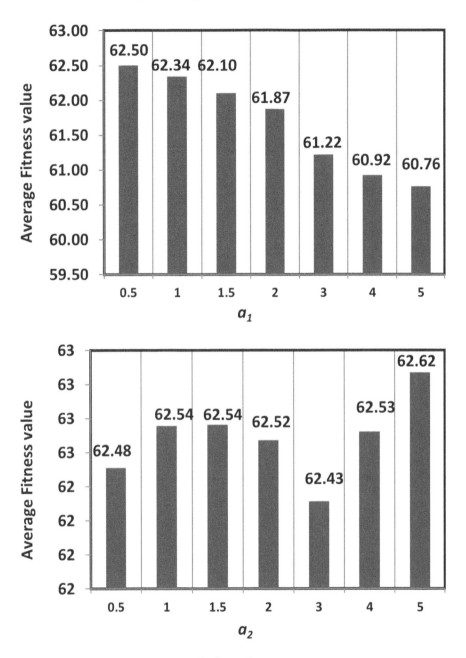

FIGURE 4.6   Tuning the parameters of EO over R1.

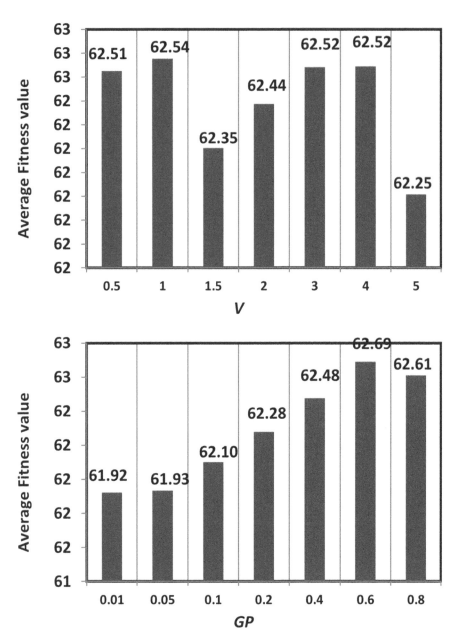

FIGURE 4.6   (Continued)

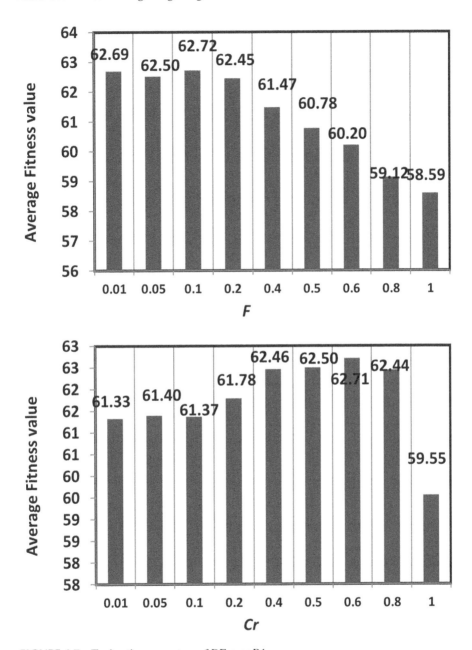

FIGURE 4.7   Tuning the parameters of DE over R1.

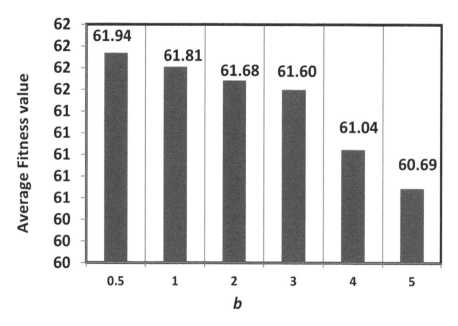

FIGURE 4.8   Tuning the parameter b of WOA over R1.

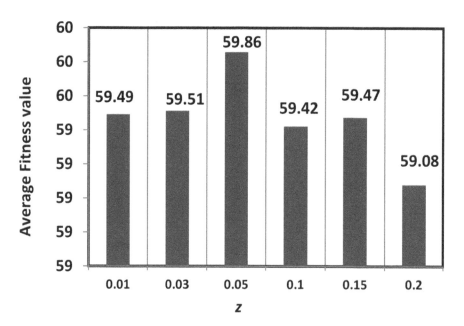

FIGURE 4.9   Tuning the parameter z of SMA over R1.

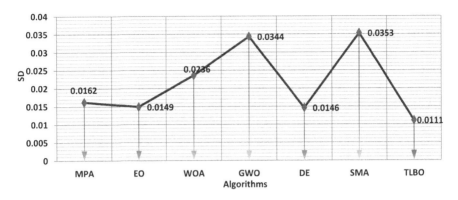

FIGURE 4.10   Comparison the average of SD values obtained by each algorithm.

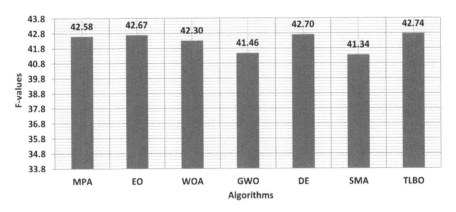

FIGURE 4.11   The average fitness values under all threshold levels for all test images.

4.3.2   COMPARISON UNDER FITNESS VALUE

The average fitness values produced by each algorithm across 30 independent runs at all threshold levels on all test images are shown in Figure 4.11. According to this figure, TLBO achieved the first position with a value of 42.74, DE achieved the second position with a value of 42.70, and EO achieved the third position with a value of 42.67, while SMA comes in the last rank with a value of 41.34. From that, it is concluded that TLBO is the top performance in terms of the objective value, while SMA is the worst.

4.3.3   COMPARISON UNDER PSNR VALUES

The predicted image quality produced by various MHAs will be compared using the PSNR metric in this section. The average PSNR values on all threshold levels for all test images within 30 runs and the averages for all levels on all photos are presented in Figure 4.12. This figure shows that DE achieves the best value compared to the

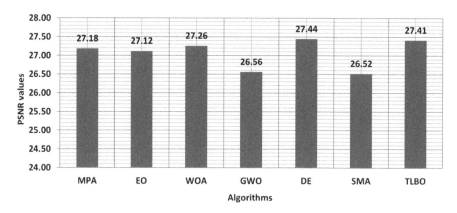

FIGURE 4.12 The average PSNR values under all threshold levels for all test images.

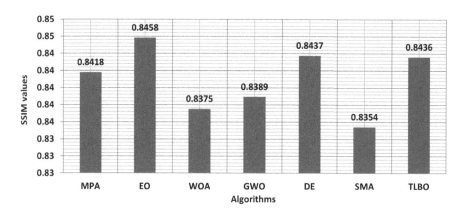

FIGURE 4.13 The average SSIM values under all threshold levels for all test images.

others and ranks first with a value of 27.44, followed by TLBO, which ranks second with a value of 27.41, and WOA, which ranks third with a value of 27.26. Meanwhile, SMA is the worst one for the PSNR metric. Finally, it is concluded that DE can reach threshold values that accurately separate the objects in an image compared to the other algorithms, and so is a strong optimizer for tackling the image segmentation problem.

4.3.4 Comparison under SSIM Values

In this section, the mean of the SSIM values generated by each MHA throughout 30 separate runs on all threshold levels under all of the medical images will be explored in further detail to demonstrate that any algorithm can achieve higher accuracy than the others. According to Figure 4.13, with a value of 0.8458, EO appears to be the greatest option, with the second-best option being DE with 0.9318, and SMA (with

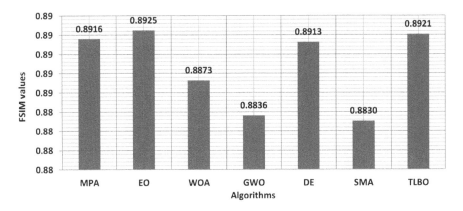

FIGURE 4.14 The average FSIM values under all threshold levels for all test images.

a value of 0.9300) appearing to be the least desirable option. SSIM is more effective than all the others because it measures the quality of the segmented images in terms of similarity, contrast distortion, and brightness. Therefore, EO is the top-performance optimizer for segmenting the medical images for improving patient healthcare. In addition, to further affirm that EO is the best, in the net section, the quality of the segmented images will be compared using another well-known metric called FSIM.

4.3.5 COMPARISON UNDER FSIM VALUES

The mean of the FSIM values obtained by each MHA across 30 successive runs at all threshold levels for all medical images will be examined in greater detail in this section to demonstrate that any method may achieve a higher level of accuracy than the others. According to Figure 4.14, EO also appears to be the best with a value of 0.8925, followed by TLBO with a value of 0.8921 and MPA with a value of 0.8916. As mentioned before, SSIM and FSIM are more effective than the other metric because they evaluate the similarity, contrast distortion, and brightness of the segmented images. As a result, EO is the best performer for segmenting medical images to improve patient care.

4.3.6 COMPARISON UNDER INTERVAL PLOT

The boxplot is utilized in this part to evaluate the algorithms' performance against fitness values on each image using threshold levels (K) of 30 and 40. To be precise, each algorithm is run 30 times on some test images with threshold levels (K) of 30 and 40, and the fitness values for these runs are plotted in Figures 4.15 to 4.29 for this algorithm. These figures show that EO's performance is so near that of TLBO, DE, and MPA, while it is much better than SMA, WOA, and GWO, which have poor performance. Finally, in the previous section, it has been clarified that EO has strong performance in terms of two more effective metrics: FSIM and SSIM, and a comparison under the boxplot of fitness values shows that it is competitive with the other

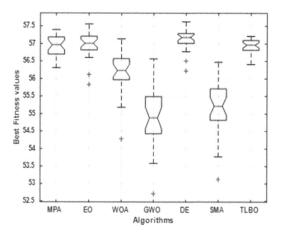

FIGURE 4.15   Fitness values on R1 image under T=30.

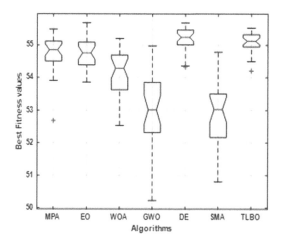

FIGURE 4.16   Fitness values on R2 image under T=30.

algorithms. Therefore, even now, EO is the best MHA for finding similar regions within a medical image in the healthcare system.

4.4   CHAPTER SUMMARY

This chapter discusses the role of metaheuristic algorithms for tackling the image segmentation problem of medical chest X-ray images in healthcare for extracting the regions which may contain features of COVID-19. MHAs have a strong perform-ance for overcoming this problem, and to show that practically, in this chapter, we selected seven well-known MHAs, which are selected based on into factors: recently published and highly cited. Those algorithms are WOA, DE, MPA, GWO, TLBO,

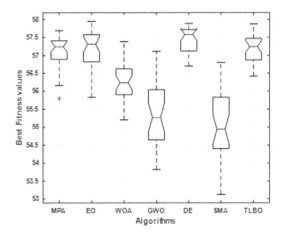

FIGURE 4.17   Fitness values on R3 image under T=30.

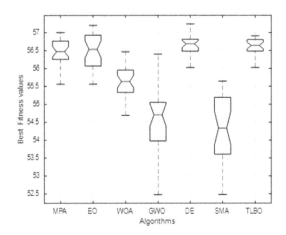

FIGURE 4.18   Fitness values on R4 image under T=30.

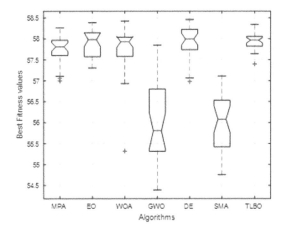

FIGURE 4.19   Fitness values on R5 image under T=30.

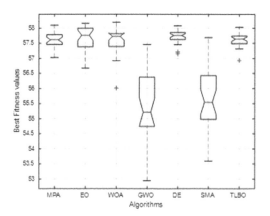

FIGURE 4.20 Fitness values on R6 image under T=30.

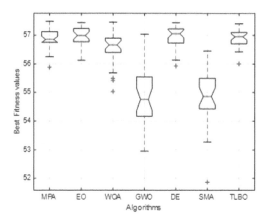

FIGURE 4.21 Fitness values on R7 image under T=30.

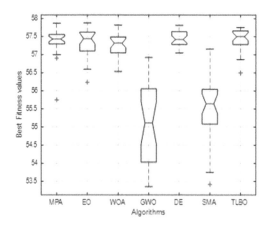

FIGURE 4.22 Fitness values on R8 image under T=30.

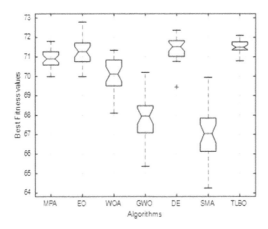

FIGURE 4.23 Fitness values on R1 image under T=40.

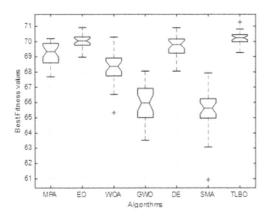

FIGURE 4.24 Fitness values on R2 image under T=40.

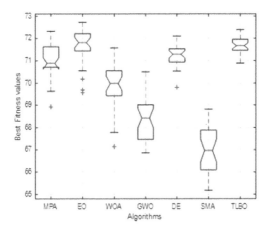

FIGURE 4.25 Fitness values on R3 image under T=40.

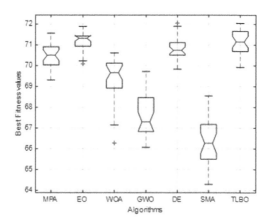

FIGURE 4.26   Fitness values on R4 image under T=40.

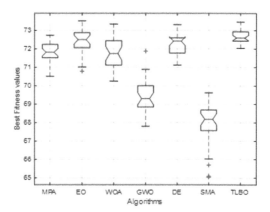

FIGURE 4.27   Fitness values on R5 image under T=40.

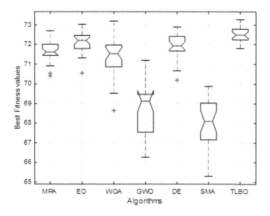

FIGURE 4.28   Fitness values on R6 image under T=40.

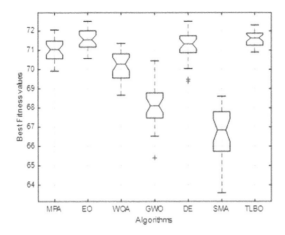

FIGURE 4.29   Fitness values on R7 image under T=40.

SMA, and EO. After that, these algorithms were validated using eight chest X-ray images under various threshold levels ranging between 3 and 40. The algorithms were compared in terms of PSNR, SSIM, FSIM, F-value, and SD. Finally, the experimental findings show that TLBO has better performance in terms of F-value and SD, and DE is better for PSNR, while EO is the top performance one in terms of SSIM and FSIM. SSIM and FSIM are more effective than the other metrics since they examine the segmented images' similarity, contrast distortion, and brightness. As a result, EO is the best segmentation algorithm for medical images to enhance patient care.

4.5   EXERCISES

a.   What is a multilevel thresholding image segmentation? Provide its pros and cons while segmenting images or extracting features

b.   How do you categorize the region-based segmentation of an image?

c.   How do you categorize the clustering-based segmentation of an image? How are they different to each other?

d.   Provide an example where a metaheuristic algorithm has excelled in image segmentation performance for a healthcare system.

e.   How do you adopt a metaheuristic model to the image segmentation problem? Provide an example of this.

f.   Provide a few names for image segmentation datasets. Demonstrate how they are varied and how they can be applied to validate a metaheuristic algorithm's performance.

g.   Why has the EO algorithm been proven to be the best segmentation algorithm? Provide some reasoning for that.

REFERENCES

1. Khan, W., Image segmentation techniques: A survey. *Journal of Image and Graphics,* 2013. **1**(4): pp. 166–170.
2. Faramarzi, A., et al., Equilibrium optimiser: A novel optimisation algorithm. *Knowledge-Based Systems,* 2020. **191**: p. 105190.
3. Abdel-Basset, M., et al., A hybrid COVID-19 detection model using an improved marine predators algorithm and a ranking-based diversity reduction strategy. *IEEE Access,* 2020. **8**: pp. 79521–79540.
4. Zhou, Y., et al., Meta-heuristic moth swarm algorithm for multi-level thresholding image segmentation. *Multimedia Tools and Applications,* 2018. **77**(18): pp. 23699–23727.
5. Upadhyay, P. and J.K. Chhabra, Kapur's entropy based optimal multi-level image segmentation using crow search algorithm. *Applied Soft Computing,* 2020. **97**: p. 105522.
6. Sharma, A., et al., Multi-level image thresholding based on Kapur and Tsallis entropy using firefly algorithm. *Journal of Interdisciplinary Mathematics,* 2020. **23**(2): pp. 563–571.
7. Küçükuğurlu, B. and E. Gedikli, Symbiotic organisms search algorithm for multi-level thresholding of images. *Expert Systems with Applications,* 2020. **147**: p. 113210.
8. Naji Alwerfali, H.S., et al., Multi-level image thresholding based on modified spherical search optimiser and fuzzy entropy. *Entropy,* 2020. **22**(3): p. 328.
9. Mahajan, S., et al., Image segmentation using multi-level thresholding based on type II fuzzy entropy and marine predators algorithm. *Multimedia Tools and Applications,* 2021. **80**(13): pp. 19335–19359.
10. Song, S., H. Jia, and J.J.E. Ma, A chaotic electromagnetic field optimisation algorithm based on fuzzy entropy for multi-level thresholding color image segmentation. *Entropy,* 2019. **21**(4): p. 398.
11. Al-Rahlawee, A.T.H., J.J.M.T. Rahebi, Applications, multi-level thresholding of images with improved Otsu thresholding by black widow optimisation algorithm. *Multimedia Tools and Applications,* 2021. **80**(18): pp. 28217–28243.
12. Khairuzzaman, A.K.M. and S.. Chaudhury, Multi-level thresholding using grey wolf optimiser for image segmentation. *Expert Systems with Applications,* 2017. **86**: pp. 64–76.
13. Sri Madhava Raja, N., et al., Otsu based optimal multi-level image thresholding using firefly algorithm. *Modelling and Simulation in Engineering,* 2014. **2014**.
14. Kapur, J.N., P.K. Sahoo, and A.K. Wong, A new method for gray-level picture thresholding using the entropy of the histogram. *Computer Vision, Graphics, and Image Processing,* 1985. **29**(3): pp. 273–285.
15. Hore, A. and D. Ziou. Image quality metrics: PSNR vs. SSIM. In *2010 20th International Conference on Pattern Recognition.* 2010. IEEE.
16. Zhang, L., et al., FSIM: A feature similarity index for image quality assessment. *IEEE Transactions on Image Processing,* 2011. **20**(8): pp. 2378–2386.

5 Role of Advanced Metaheuristics for DNA Fragment Assembly Problem

5.1 DNA FRAGMENT ASSEMBLY PROBLEM

DNA, also known as deoxyribonucleic acid, is the genetic material that holds the information necessary to generate all alive organisms. It is made up of four different types of chemical bases, namely adenine (A), cytosine (C), thymine (T), and guanine (G). The order in which these four letters (A, G, C, and T) appear reveals the information accessible for constructing the live organism. Base pairs are formed when letters combine in pairs, such as when A is combined with T or when C is combined with G. To construct a nucleotide, each pair is joined to a sugar molecule and a phosphate molecule. The structure of a nucleotide is analogous to that of a ladder, with two long strands that wind around one another to form a double helix [1, 2].

In the discipline of computational biology, these sequences are utilized to determine the function of DNA-coded information. The human genome has a large number of bases (about 3.2 billion), yet current sequencing technology cannot read more than 1,000 bases. To circumvent this issue, a shotgun sequencing approach has been employed to divide the lengthy DNA sequence into fragments or segments. These fragments are sequenced at random by a machine, destroying their original order and orientation. To recover the original order of the DNA, the fragments must be reconstructed based on the overlap between them. The assembler must compute the overlap between all possible pairings of fragments to reassemble the fragments with the highest similarity score into contigs. The contig consists of elements that are contiguous and overlap. The task is to reassemble the contigs to recover the original DNA successfully.

The classic assembly method consists of three stages: overlap, layout, and consensus. This method is referred to as the overlap-layout-consensus (OLC) technique. During the overlapping stage, the assembler determines the percentage of overlap between the available fragments. This stage aims to identify the fragment with the highest overlap score between the prefix of one fragment and the suffix of another fragment. The semi-global alignment approach is selected, and dynamic programming is used to implement it [3–5]. In the stage known as "layout," the order of the fragments is reassembled in such a way as to maximize the sum of all the overlaps produced by each adjusted fragment. This process continues until the initial DNA sequence is obtained. In the final stage, known as the consensus step, the order of

DOI: 10.1201/9781003325574-5

the fragments that were determined in the previous step, known as the layout step, is utilized to build the entire DNA. DNA research has enabled the early diagnosis and prediction of an individual's susceptibility to numerous diseases, including cancer [6–8] and autoimmune [9] diseases.

DNA strands are typically divided into short pieces or fragments at random locations to assist the reading process. Upon completion of the study, the DNA fragments must be recombined into the original DNA sequence; this is known as the DNA fragment assembly problem (DFAP). In DFAP, the primary purpose is to discover the ideal arrangement of fragments to reassemble the original DNA sequence; consequently, the layout phase is considered the heart of DFAP. To guarantee the precision of any reassembled DNA, the two primary goals of DFAP must be achieved: (1) to reach the optimal order of the fragments that combine the original DNA; and (2) simultaneously optimizing the overlap score among these fragments and reducing the total number of contigs. The conventional methods of DFAP examine all conceivable fragment combinations to identify the most effective one. There are $2^F F!$ potential combinations for F fragments, causing the solution time to expand exponentially, may consume years to find the exact solution [3]. Several heuristic and metaheuristic methods have been presented to solve the DFAP to provide near-optimal solutions in an acceptable amount of processing time to overcome these limitations. Some of the heuristic an metaheuristic proposed for DFAP are: problem aware local search (PALS) [10], genetic algorithm (GA) [4, 11], particle swarm optimization (PSO) [12, 13], firefly algorithm (FA) [14], ant colony optimization algorithm (ACO) [15], harmony search algorithm (HAS) [16], gravitational search algorithm (GSA) [17], bee algorithm (BA) [18], cuckoo search algorithm (CSA) [19], and whale optimization algorithm (WOA) [20].

This chapter will begin with a discussion of how to design metaheuristic algorithms to solve this problem generally. After that, the choice of control parameters and the experimental settings to solve DFAP are explained. In the end, but certainly not least, a few experiments are carried out to determine which algorithm would be the most effective in solving this problem.

5.2   METAHEURISTIC-BASED DNA FRAGMENT ASSEMBLY PROBLEM (DFAP)

DFAP is a combinatorial problem that aims to enhance the overlap between adjacent fragments to construct a single contig of DNA. A thorough comprehension of this issue is required to create a good solution illustration. The fragments are counted from 1 to F, where F is the maximum number of fragments. Then, one potential solution to DFAP is to reorder the numbers from 1 to F. Examining the permutations of the fragments' given numbers is necessary for determining the ideal arrangement of the fragments. Now, to solve this DFAP using a metaheuristic method, a population of N solutions will be created such that each one represents a solution to this problem. Each solution is described by a position vector of size F, including a permutation of numbers allocated to the fragments. Each cell within the solution position vector must have a unique fragment number. To further elaborate on the solution representation

| 1 | 5 | 4 | 3 | 2 | 6 | 8 | 7 | 10 | 9 |

FIGURE 5.1   Representation of a DFAP solution $(x_1, F = 10)$.

| 5 | 4 | 3 | 1 | 2 |

| 2 | 3 | 4 | 1 | 5 |

FIGURE 5.2   Two solutions generated by a metaheuristic algorithm $(x_1, F = 5)$.

of this problem, Figure 5.1 is presented to depict how to set the values of a solution that is applicable for solving the DFAP. In this figure, we set the value of F to 10, then create a vector including F cells which are initially assigned integers generated randomly between 1 and 10; these integers represent the fragment id, such that 1 represents the first fragment, 2 indicates the second one, and so on.

After initializing the population, the quality of the solutions must be computed to determine the optimal solution for this problem. To do this, an objective function is necessary. In this problem, the objective function determines the fitness value of each population solution. The solution with the highest fitness value is determined to be the optimal solution. The primary purpose of DFAP is to achieve the best arrangement of fragments, which maximizes the overlap score and lowers the number of contigs. To compute the fitness value of each solution in order to locate the solution closest to the ideal one, Equation 5.1 is considered.

$$F\left(\overrightarrow{X_i}(t)\right) = \sum_{d=0}^{N-2} w\left(f_d, f_{d+1}\right)$$ (5.1)

where $F\left(\overrightarrow{X_i}(t)\right)$ indicates the fitness values of the ith solution, and $w\left(f_d, f_{d+1}\right)$ stands for the overlap score between any two successive fragments. As identified previously, the overlap score has been calculated using semi-global alignment, implemented by a dynamic programming approach. For example, suppose that there are five fragments with an overlap score between every two fragments shown in Table 5.1, and the metaheuristic algorithms generated two solutions depicted in Figure 5.2. Afterwards, the objective function (Equation 5.1) will be applied to those solutions to determine the quality of each one as follow:

Step 1: Calculating the overlap score between each two successive fragment in each solution:

X_1: $w(X_{1,0}, X_{1,1})$= w(5, 4)=22, according to the overlap score matrix
X_1: $w(X_{1,1}, X_{1,2})$= w(4, 3)=17, according to the overlap score matrix
X_1: $w(X_{1,2}, X_{1,3})$= w(3, 1)=10, according to the overlap score matrix
X_1: $w(X_{1,3}, X_{1,4})$= w(1, 2)=6, according to the overlap score matrix

TABLE 5.1
Overlap score matrix

Fragment ID	1	2	3	4	5
1	0	6	7	4	10
2	4	0	3	7	8
3	10	11	0	12	13
4	14	15	17	0	18
5	19	20	21	22	0

10	12	5.2	7.3	1	2.3	0.5	-0.7	1.5	-30

FIGURE 5.3 Continuous solution produced by a metaheuristic algorithm $(x_1, F = 10)$.

X_2: $w(X_{1,0}, X_{1,1})$ = $w(2, 3)$=3, according to the overlap score matrix
X_2: $w(X_{1,1}, X_{1,2})$ = $w(3, 4)$=12, according to the overlap score matrix
X_2: $w(X_{1,2}, X_{1,3})$ = $w(4, 1)$=14, according to the overlap score matrix
X_2: $w(X_{1,3}, X_{1,4})$ = $w(1, 5)$=10, according to the overlap score matrix

Step 2: Computing the overall fitness value of each solution as follows:

$$F\left(X_1\right) = 22+17+10+6 = 55$$
$$F\left(X_2\right) = 3+12+14+10 = 39$$

Because $F\left(X_1\right) > F\left(X_2\right)$ and this problem is a maximization problem, X_1 is extracted as the best-so-far solution to guide the other solutions during the optimization process.

Unfortunately, the typical metaheuristic algorithms are created to handle problems with continuous-search space and, therefore, cannot be used in the discrete-search space of the DFAP. In the literature, authors provide mapping approaches such as the smallest position value (SPV) or the largest position value (LPV) for transforming continuous optimization-based algorithms to discrete ones. SPV sorts the continuous values of the solutions in ascending order so that the smallest position value is mapped to 1; the next smallest position value is mapped to 2, and so on. LPV, unlike SPV, arranges the solutions' continuous values in descending order. The greatest position value is assigned the value 1. Likewise, the value of the second-largest location is mapped to 2, and so on. In this chapter, we employ SPV to convert the continuous values of the investigated metaheuristic algorithms to discrete values applicable to the DFAP. For instance, supposing that a metaheuristic algorithm like WOA generates the continuous solutions represented in Figure 5.3, this solution does not apply to DFAP since it needs a combinatorial solution, not a continuous one. Therefore, the SPV strategy is applied to make this solution applicable to this problem by sorting these continuous values in ascending order according to the rank of each value in

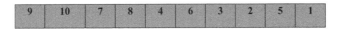

FIGURE 5.4 Rank of each value in the continuous X_l $(x_1, F = 10)$.

TABLE 5.2
Description of DFAP benchmark instances

ID	Instances	Abbreviation	Coverage	AFL	NF	OSL
		GenFrag				
1	X60189(4)	X1	4	395	39	3835
2	X60189(5)	X2	5	286	48	3835
3	X60189(6)	X3	6	286	48	3835
4	X60189(7)	X4	7	387	68	3835
5	M15421(5)	M1	5	398	127	10089
6	M15421(6)	M2	6	350	173	10089
7	M15421(7)	M3	7	383	177	10089
8	J02459(7)	J1	7	405	352	20000
		F-series				
9	F25(305)	F1	–	307	25	7630
10	F25(400)	F2	–	400	25	10006
11	F25(500)	F3	–	500	27	13051
12	F50(315)	F4	–	315	50	15791
13	F50(412)	F5	–	412	50	20628
14	F50(498)	F6	–	498	50	24956
15	F100(307)	F7	–	307	100	30443
16	F100(415)	F8	–	415	100	–
17	F100(512)	F9	–	512	100	–
18	F508(354)	F10	–	354	508	–
19	F635(350)	F11	–	350	635	–
20	F737(355)	F12	–	355	737	–
21	F1343(354)	F13	–	354	1343	–

the continuous solution, as shown in Figure 5.4, such that –30 is the smallest value, is given a rank value of 1, and the second smallest one which has a value of –0.7 in Figure 5.3 is given a value of 2, and so on.

5.3 EXPERIMENT SETTINGS

Several tests have been undertaken to evaluate the effectiveness of seven previously described metaheuristic algorithms: WOA, GWO, SMA, EO, DE, TLBO, and MPA, using 21 benchmark instances. Twenty-one instances are derived from two benchmarks [21]: GenFrag consists of 8 instances, while f-series contains 13. According to [21], Table 5.2 describes the 21 instances in terms of coverage, a number of fragments

(NF), average fragment length (AFL), and original sequence length (OSL). Coverage can be estimated using Eq. (5.2) [1].

$$Coverage = \frac{\sum_{j=1}^{NF} length\,of\,fragment\,j}{length\,of\,target\,fragment} \qquad (5.2)$$

To ensure fragment overlap throughout the reassembly process, the coverage value must be greater than 1. In Table 5.2, known coverage is greater than 1, AFL varies between 300 and 600 bases, whereas NF ranges between 25 and 1343. All the studies are conducted on Windows 10 with 32GB RAM and a 2.40 GHz Intel Core i7-4700MQ processor. This chapter's algorithms are all implemented using the Java programming language under a population size and a maximum number of iterations of 30 and 5000, respectively. In addition, statistical analyses are employed to validate the results.

5.4   CHOICE OF PARAMETERS

Regarding the controlling parameters of the analysed metaheuristic algorithms (MHAs), this chapter presents the results of comprehensive experiments designed to identify the parameter value that is most appropriate for each algorithm in terms of its ability to solve the DFAP. Extensive experiments under various values for each parameter have been conducted to achieve this goal for MPA, which has two primary parameters: P and FADs, which need to be accurately estimated to maximize its performance. The average fitness values that have been obtained under each of the various values for each parameter are depicted in Figure 5.5. Taking a closer look at this figure reveals that the optimal value for both P and FADs is 4 and 0.3, respectively. To get the most out of the EO method, its performance can be improved by precisely estimating its four primary effective parameters, denoted by the letters a_1, a_2, V, and GP. As a result, many different values for these parameters are being researched to determine which improves EO's overall performance (see Figure 5.6). From Figure 5.6, we may conclude that the optimal values for these parameters are, in order, 1, 2.5, 1, and 0.8. When it comes to the parameters of DE, which are F and Cr, the optimal values for them are estimated by running DE under a variety of different values for each parameter, and the results that were achieved (the average fitness value for each value of each parameter) are displayed in Figure 5.7. Based on the results shown in this figure, we determined that 0.3 and 0.02 were the best possible values for F and CR, respectively.

Last but not least, WOA has one effective parameter, namely b. This parameter is responsible for defining the shape of the logarithmic spiral. Tests have been conducted under various values, such as 0.5, 1, 2, 3, 4, and 5, to determine the optimal value for this parameter. The results of these experiments demonstrate that the optimal value for this parameter is 0.5, as shown in Figure 5.8. Finally, the SMA has one effective parameter denoted by the letter z. This parameter's effectiveness is assessed using the data presented in Figure 5.9, which was derived from a large number of tests in which

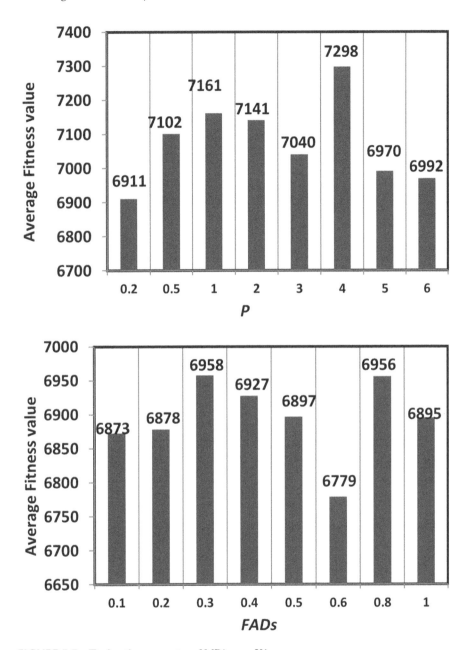

FIGURE 5.5 Tuning the parameter of MPA over X1.

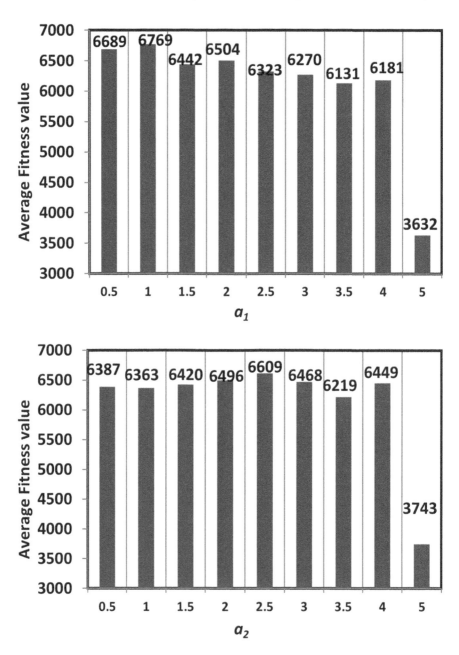

FIGURE 5.6 Tuning the parameters of EO over X1.

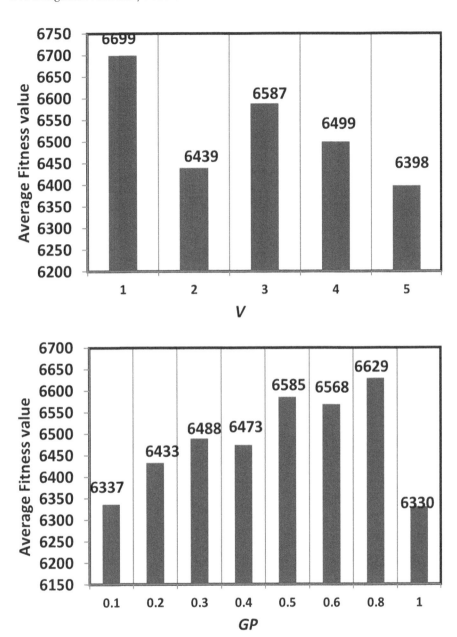

FIGURE 5.6 (Continued)

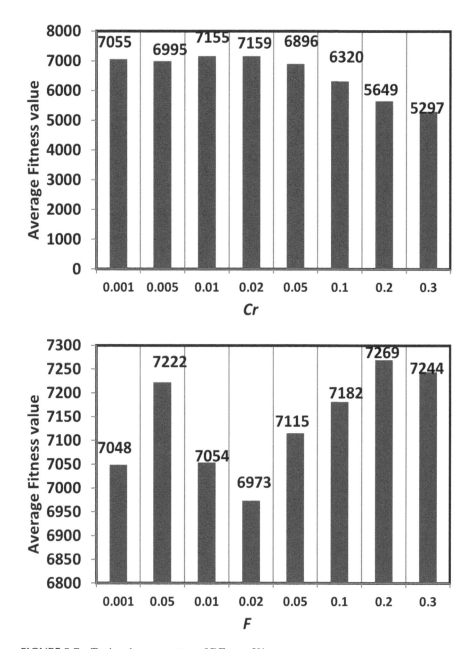

FIGURE 5.7 Tuning the parameters of DE over X1.

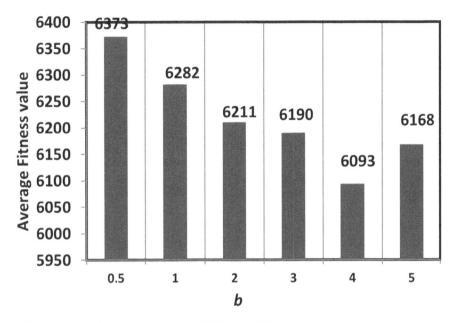

FIGURE 5.8   Tuning the parameter b of WOA over X1.

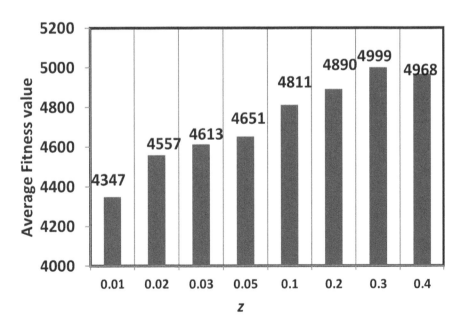

FIGURE 5.9   Tuning the parameter z of SMA over X1.

different values were used for this parameter. The value of 0.3 seems to be the optimal choice for this parameter, according to this figure.

5.5   RESULTS AND DISCUSSION

In the beginning, each algorithm is carried out for a total of 20 separate runs. The results of these runs, including the average overlap score and the average number of contigs, are reported in Tables 5.3 and 5.4, respectively. After that, those results were analysed to determine how well each algorithm worked under such conditions. This observation demonstrates that, in terms of the average fitness value, which is also referred to as the average overlap score, MPA may be superior to the other algorithms for the large-scale instance that has a number of fragments that is greater than 100, whereas DE may be more effective for the other instances that have a number of fragments less than that. In a similar vein, in terms of the average number of contigs obtained by each method, MPA may be superior to the other algorithms for the large-scale instance, whilst DE may be superior for the other instance.

Figure 5.10 is a graphical representation of the average results found in Table 5.3 for each algorithm. The purpose of this representation is to make the efficiency of each algorithm more readily apparent. This figure demonstrates that the MPA method

TABLE 5.3
A comparison of average fitness among various algorithms

Instance	MPA	EO	WOA	GWO	DE	SMA	TLBO
X4	6831.8	6345.45	6271.35	5134.65	**7206.45**	4467.2	5514.7
X5	8631.25	8023.1	7708.0	6291.05	**8864.45**	5434.25	6725.95
X6	9536.7	8767.75	8357.1	6601.25	**9564.55**	5969.45	7064.25
X7	**11568.85**	10572.75	10198.75	8312.6	11381.5	7318.95	8599.95
M5	**13169.45**	10154.45	10134.3	7601.35	10443.45	6289.4	7699.3
M6	**13720.25**	10757.0	10428.3	7905.45	9652.65	6728.65	7805.0
M7	**16310.75**	12477.4	12466.5	9183.0	11428.3	7779.75	9342.3
J7	**23464.55**	15652.6	15484.45	11903.75	12974.8	10516.15	11962.85
F305	516.2	495.35	492.35	429.7	**533.4**	405.7	493.9
F400	642.35	635.45	610.35	512.85	**665.1**	474.15	621.3
F500	764.65	745.9	712.6	586.5	**778.35**	523.9	703.0
F315	1047.05	989.7	940.2	785.35	**1081.35**	706.8	847.1
F412	1033.7	981.3	936.1	788.1	**1065.1**	724.35	826.5
F498	1050.75	985.1	962.25	809.3	**1072.95**	730.25	856.1
F307	**1575.5**	1450.75	1388.15	1180.25	1443.1	1094.6	1223.45
F415	**1598.05**	1442.1	1370.15	1177.5	1439.0	1112.45	1225.3
F512	**1582.0**	1437.25	1354.2	1186.2	1425.35	1098.85	1218.3
F508	**5655.55**	4952.35	4911.35	4578.9	4645.75	4460.05	4589.4
F635	**6750.9**	5931.5	5906.45	5634.95	5644.6	5508.8	5616.65
F737	**7511.75**	6671.65	6698.55	6412.55	6428.6	6293.7	6415.25
F1343	**12890.35**	11653.1	11751.7	11435.25	11396.7	11265.6	11402.85

Bold values indicate the best results

TABLE 5.4
A comparison of number of contigs among various algorithms

Instance	MPA	EO	WOA	GWO	DE	SMA	TLBO
X4	4.15	4.05	2.8	3.75	**1.55**	3.2	2.55
X5	6.1	6.6	5.4	5.65	**3.0**	6.3	4.3
X6	8.15	8.5	7.7	8.8	**5.2**	9.35	6.7
X7	7.25	7.95	5.9	6.4	**4.2**	6.7	5.05
M5	21.55	27.6	23.7	25.8	**20.85**	25.2	22.9
M6	**35.85**	44.75	41.85	45.6	39.8	46.65	42.3
M7	**34.8**	42.0	38.85	40.85	35.4	41.85	37.25
J7	**63.6**	81.6	75.2	76.2	72.1	76.25	72.35
F305	8.5	9.85	9.75	12.6	**6.6**	12.35	8.65
F400	8.9	9.55	9.35	12.0	**7.1**	12.8	8.95
F500	9.4	10.6	10.15	12.4	**8.0**	13.55	10.05
F315	20.05	21.95	20.75	24.85	**15.6**	26.95	22.9
F412	16.75	19.2	17.4	22.2	**13.85**	22.9	19.5
F498	16.6	18.65	18.0	22.45	**13.65**	23.6	19.3
F307	**41.55**	48.05	46.8	56.4	43.25	58.3	53.3
F415	**40.8**	44.8	44.75	54.65	42.3	57.35	51.6
F512	**38.7**	45.2	46.45	54.6	41.35	57.65	51.0
F508	**295.7**	342.35	340.2	360.7	353.6	366.25	353.8
F635	**376.0**	432.35	430.7	454.0	446.2	460.1	447.8
F737	**443.65**	514.3	511.5	533.2	523.25	539.75	524.2
F1343	**851.3**	968.85	956.1	979.2	974.75	992.95	973.8

Bold values indicate the best results

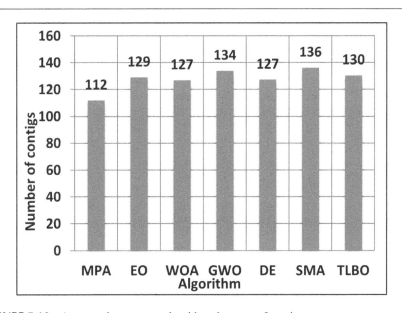

FIGURE 5.10 A comparison among algorithms in terms of contigs.

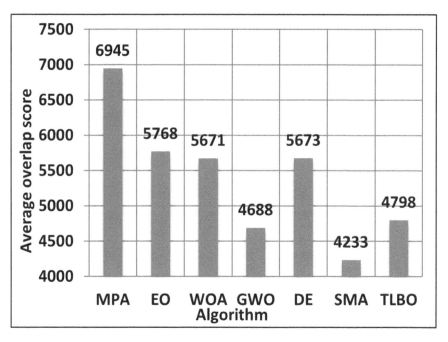

FIGURE 5.11   A comparison among algorithms in terms of overlap score.

performs the best among the other algorithms with a value of 6945, while the SMA algorithm performs the worst with an amount of 4233. The average results that were reported in Table 5.4 for each method are graphically depicted in Figure 5.11 to make it easier to understand which approach had the potential to reach the fewest number of contigs. This chart demonstrates that the MPA algorithm performs better than the other algorithms with a value of 112, while the SMA algorithm performs the worst with a quantity of 136. In addition, Figure 5.12 presents a measurement of the outcomes' distribution based on five metrics: the minimum, the first quartile (Q2), the median, the third quartile (Q3), and the maximum for the examples X1, X2, X3, X4, M1, M2, M3, and J1. Once more, this figure demonstrates that the MPA is superior in terms of X3, X4, M1, M2, M3, and J1. When there are a greater number of fragments, the MPA method performs significantly better than its competitors.

5.6   CHAPTER SUMMARY

This chapter investigates the performance of seven well-established metaheuristic algorithms, namely WOA, GWO, MPA, TLBO, SMA, EO, and DE, for the DNA fragment assembly problem to aid in the early diagnosis and prediction of an individual's susceptibility to numerous diseases. These metaheuristic algorithms were designed for continuous-search space and cannot be applied in DFAP's discrete-search environment. In the literature, authors use SPV or LPV to transfer continuous optimization-based techniques to discrete ones. SPV arranges the solutions'

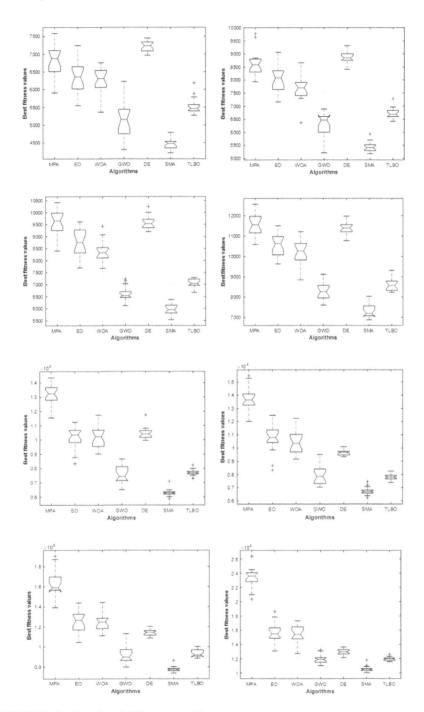

FIGURE 5.12 Boxplot of different algorithms.

continuous values in ascending order, so the smallest position value is mapped to 1, etc. LPV sorts continuous values in decreasing order, unlike SPV. This chapter employs SPV to transform continuous metaheuristic algorithm values into discrete DFAP values. Several tests have been done to evaluate seven previously disclosed metaheuristic algorithms utilizing 21 benchmark instances. The experimental results and statistical analyses show that MPA outperforms the other algorithms significantly in reaching less number of contigs, and an average overlap score for the instances with a number of fragments greater than 100. For the other instances, MPA is competitive with DE. Generally, MPA is a strong alternative to tackle this problem for reaching the near-optimal order of the fragments, which significantly affects the original DNA sequence.

5.7 EXERCISES

a. What is an overlap layout consensus technique? What is the main purpose of using this technique?
b. Define the DNA fragment assembly problem. Provide an example where a metaheuristics algorithm has proven effective for DNA fragment assembly problems.
c. What are the traditional ways to tune algorithmic parameters for metaheuristics? How can we customize them for the DNA fragment assembly problem?
d. How can we adopt multiple metaheuristics to solve DNA fragment assembly problems?
e. Why do we need advanced optimization techniques for the DNA fragment assembly problem?

REFERENCES

1. Allaoui, M., B. Ahiod, and M. El Yafrani, A hybrid crow search algorithm for solving the DNA fragment assembly problem. *Expert Systems with Applications,* 2018. **102**: pp. 44–56.
2. Zhou, Z., B. Yang, and W. Hou, Association classification algorithm based on structure sequence in protein secondary structure prediction. *Expert Systems with Applications,* 2010. **37**(9): pp. 6381–6389.
3. Kubalik, J., P. Buryan, and L. Wagner. Solving the DNA fragment assembly problem efficiently using iterative optimization with evolved hypermutations. In *Proceedings of the 12th annual conference on Genetic and evolutionary computation.* 2010. ACM.
4. Nebro, A.J., et al., DNA fragment assembly using a grid-based genetic algorithm. *Computers & Operations Research,* 2008. **35**(9): pp. 2776–2790.
5. Patra, S., S. Goswami, and B. Goswami, Fuzzy and simulated annealing based dynamic programming for the unit commitment problem. *Expert Systems with Applications,* 2009. **36**(3): pp. 5081–5086.
6. Phillips-Wren, G., P. Sharkey, and S.M. Dy, Mining lung cancer patient data to assess healthcare resource utilization. *Expert Systems with Applications,* 2008. **35**(4): pp. 1611–1619.
7. Maleki, N., Y. Zeinali, and S.T.A. Niaki, A k-NN Method for lung cancer prognosis with the use of a genetic algorithm for feature selection. *Expert Systems with Applications,* 2020: p. 113981.

8. Wang, Q., et al., Adaptive sampling using self-paced learning for imbalanced cancer data pre-diagnosis. *Expert Systems with Applications,* 2020. **152**: p. 113334.

9. Lo, Y.-M.D., et al., *Sequencing analysis of circulating DNA to detect and monitor autoimmune diseases.* 2019, Google Patents.

10. Alba, E. and G. Luque. A new local search algorithm for the DNA fragment assembly problem. In *European Conference on Evolutionary Computation in Combinatorial Optimization.* 2007. Springer.

11. Hughes, J.A., S. Houghten, and D. Ashlock, Restarting and recentering genetic algorithm variations for DNA fragment assembly: The necessity of a multi-strategy approach. *Biosystems,* 2016. **150**: pp. 35–45.

12. Rajagopal, I. and U. Maheswari Sankareswaran, An adaptive particle swarm optimization algorithm for solving DNA fragment assembly problem. *Current Bioinformatics,* 2015. **10**(1): pp. 97–105.

13. Huang, K.-W., et al., A memetic particle swarm optimization algorithm for solving the DNA fragment assembly problem. *Neural Computing and Applications,* 2015. **26**(3): pp. 495–506.

14. Ezzeddine, A.B., S. Kasala, and P. Navrat. Applying the firefly approach to the DNA fragments assembly problem. In *Annales Univ. Sci. Budapest., Sect. Comp.* 2014.

15. Meksangsouy, P. and N. Chaiyaratana. DNA fragment assembly using an ant colony system algorithm. In *The 2003 Congress on Evolutionary Computation, 2003. CEC'03.* 2003. IEEE.

16. Ülker, E.D. Adaptation of harmony search algorithm for DNA fragment assembly problem. In *2016 SAI Computing Conference (SAI).* 2016. IEEE.

17. Huang, K.-W., et al., A memetic gravitation search algorithm for solving DNA fragment assembly problems. *Journal of Intelligent & Fuzzy Systems,* 2016. **30**(4): pp. 2245–2255.

18. Firoz, J.S., M.S. Rahman, and T.K. Saha. Bee algorithms for solving DNA fragment assembly problem with noisy and noiseless data. In *Proceedings of the 14th Annual Conference on Genetic and Evolutionary Computation.* 2012. ACM.

19. Indumathy, R., S.U. Maheswari, and G. Subashini, Nature-inspired novel Cuckoo Search Algorithm for genome sequence assembly. *Sadhana,* 2015. **40**(1): pp. 1–14.

20. Abdel-Basset, M., et al., An efficient-assembler whale optimization algorithm for DNA fragment assembly problem: Analysis and validations. *IEEE Access,* 2020. **8**: pp. 222144–222167.

21. Mallén-Fullerton, G.M., et al., Benchmark datasets for the DNA fragment assembly problem. *International Journal of Bio-Inspired Computation,* 2013. **5**(6): pp. 384–394.

6 Contribution of Metaheuristic Approaches for Feature Selection Techniques

6.1 FEATURE SELECTION PROBLEM IN HEALTHCARE

Various fields have implemented several technologies and vast amounts of data. Data mining plays a crucial role in disease prediction in the medical field. Data mining employs machine learning, artificial intelligence, and statistical prowess to create highly accurate predictive models for crucial areas, like cancer diseases and heart diseases [1]. The fast increase in the amount of data stored, processed, and retrieved by systems has made it challenging to extract critical and useful information from them [2–4]. This is because acquired data frequently contain redundant and irrelevant information that can hinder the success of an operation. Medical datasets are overloaded with irrelevant and redundant information, resulting in dimensionality that is difficult to analyse. As a result, the learning process suffers in terms of accuracy, cost, and speed. Accuracy and sensitivity are critical when it comes to medical diagnosis and treatment. Doctors will be able to identify patients more promptly and at a lower cost if they can analyse medical data more efficiently and accurately. To avoid these duplicate characteristics and their negative effects, it is vital to develop a strategy to avoid them.

Dimensionality Reduction (DR) is a technique for accomplishing this [5]. DR is a technique that eliminates unimportant, redundant, and irrelevant characteristics from datasets without diminishing the amount of data they include. DR approaches can be split into two distinct categories. The primary procedure is feature extraction (FE). The FE technique reduces the dimension of high-dimensional data by extracting new feature spaces from the original dataset's features. The next category is feature selection (FS). FS is the procedure that extracts the fewest number of the most relevant and meaningful features from the original dataset's features without losing information. It has proven to be a valuable tool for removing redundant and irrelevant features without information loss. Three types of FS approach exist: supervised [6], unsupervised [7], and semi-supervised [8]. These categories are established by the dataset's dependency degree of the class label.

Based on the level of the learning algorithm used to evaluate feature subsets, FS techniques can be split into three distinct strategies. The first strategy, referred to as the filter method, employs filters that exclude learning algorithms while selecting features that are independent of categorization and other learning tasks. The filter

DOI: 10.1201/9781003325574-6

approach prioritizes features based on the data's internal relationships. Filters are considered a quick method. Some of the well-known filter methods are chi-square, gain ratio, information gain (IG), and ReliefF. The second technique is wrappers, which employ a particular learning technique to evaluate each selected subset of features by a predictor. This strategy improves classification accuracy, but the optimization process is slow. The embedded method is the last strategy. This strategy incorporates FS into the training process to coordinate learning velocity and model performance [5].

This chapter will first review some previously proposed feature selection techniques and describe how to adapt the metaheuristic algorithms for tackling this problem generally. Then, the performance metrics and experimental settings are presented. Last but not least, some experiments are conducted to show which algorithm could tackle this problem better. Finally, a chapter summary is presented.

6.2 METAHEURISTIC-BASED FEATURE SELECTION

In the field of FS, metaheuristic algorithms have become increasingly popular as a solution to the time complexity challenge posed by classic methods such as mutual information, information gain, relief, depth search, and breadth search. In recent times, numerous metaheuristic algorithms have been offered as potential solutions to this problem, such as gradient descent algorithm (GDA) [9], binary duck travel optimization algorithm (BDTO) [10], quantum-based whale optimization algorithm (QWOA), novel chaotic crow search algorithm (CCSA) [11], improved binary sailfish optimizer (BFO) [12], novel chaotic selfish herd optimizer (CSHO) [13], fish swarm optimization (FSO) [14], chaotic dragonfly algorithm (CDFA) [15], S-shaped binary whale optimization algorithm (BWOA) [16], binary particle swarm optimization with time-varying inertia weight strategies [17], grey wolf optimizer with a two-phase mutation strategy (GWOTM) [18], and a discrete binary version of the particle swarm optimization (BPSO) [19].

The vast majority of metaheuristic algorithms begin by generating a population with the size N and a number of dimensions d for each member of the population. After this, the dimensions are initialized by the characteristics of the problems being tackled. Because FS problems are discrete, the dimensions of each solution which are equal to the length of the features contained within the datasets, will be randomly initialized with 0 and 1 to mark the selected features to identify the optimal subset of the features that could reach better classification or clustering accuracy. See Figure 6.1 for further clarification and illustration; it illustrates how to initialize the solutions when solving the FS problem.

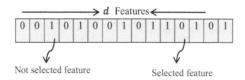

FIGURE 6.1   Representation of a feature selection solution.

After randomly initializing N solutions with binary values and evaluating each solution to determine the best-so-far one which has the lowest fitness value, the optimization process will start looking for a solution that is superior to the best-so-far one to find a better subset of selected features. However, due to the continuous nature of the problem that needs to be solved, the updated solutions produced by the metaheuristic algorithms cannot be used. Because of this, V-shaped and S-shaped transfer functions will be used to convert the continuous values produced by the metaheuristic algorithms so that they can be applied to this problem. Following this, an in-depth discussion of the various transfer functions will occur.

6.2.1 V-Shaped and S-Shaped Transfer Function

Finding a solution to the discrete FS problem with any metaheuristic approach that produces continuous values is impossible. As a consequence, the transfer function also referred to as a transformation function, has been implemented to normalize the continuous values produced by the metaheuristic algorithm within the range of 0 to 1. After the continuous values have been normalized, they are transformed into the values 0 and 1 using Equation (6.1). As shown in Table 6.1 and graphically shown in Figure 6.2, transfer functions can be classified into two different shapes: V-shaped, and S-shaped.

$$V_j = \begin{cases} 1 & if\ V_j > 0.5 \\ 0 & otherwise \end{cases} \tag{6.1}$$

6.2.2 Evaluation Phase: Objective Function

The FS is considered a multi-objective issue since it has two objectives that are in conflict: maximizing the accuracy while simultaneously lowering the feature length. However, the vast majority of metaheuristic algorithms have been proposed to address

TABLE 6.1
V-shaped and S-shaped transfer function

V-Shaped	S-Shaped
V1 $F(a) = \left\| \frac{2}{\pi} arcTan\left(\frac{\pi}{2}a\right) \right\|$	S1 $F(a) = \frac{1}{1+e^{-a}}$
V2 $F(a) = \|tanh(a)\|$	S2 $F(a) = \frac{1}{1+e^{-2*a}}$
V3 $F(a) = \left\| \frac{a}{\sqrt{1+a^2}} \right\|$	S3 $F(a) = \frac{1}{1+e^{-\frac{a}{2}}}$
V4 $F(a) = \left\| erf\left(\frac{\sqrt{\pi}}{2}a\right) \right\|$	S4 $F(a) = \frac{1}{1+e^{-\frac{a}{3}}}$

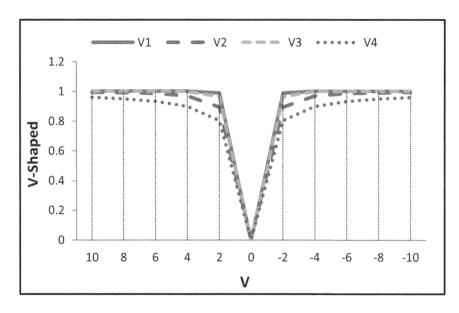

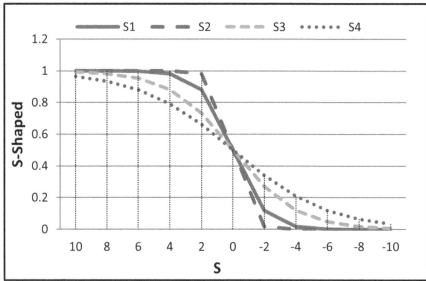

FIGURE 6.2 (a) V-shaped and (b) S-shaped transformation functions.

issues with a single objective, and as a result, they are not relevant to problems with multiple objectives. A number of studies tackled this issue in two distinct ways: the first utilized Pareto optimality, while the second utilized weighting variables to combine two objectives into a single one. This chapter uses the second method to transform the multi-objective FS into a single objective. This method makes use of a weighting variable α, which is given a value between 0 and 1 depending on the

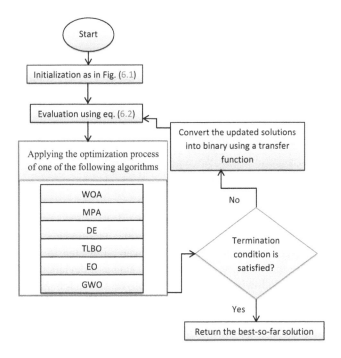

FIGURE 6.3 Flowchart of adapting metaheuristics for the feature selection problem.

degree to which one objective is preferred over the others. The proposed algorithm's objective function, which is used in general to find the smallest subset of features as the second desired objective after maximizing accuracy, is formulated as that:

$$f = \alpha * \gamma_R (D) + \beta * \frac{|S|}{|N1|} \tag{6.2}$$

Where $\gamma_R (D)$ indicates the classification error rate that was fed back from the K-nearest neighbour algorithm (KNN) using the holdout approach that was used to partition the dataset into the training dataset and the testing dataset [20, 21], $|S|$ is the subset of selected features by a metaheuristic algorithm, $|N1|$ refers to the feature number in the employed dataset, and β is of $\beta = (1 - \alpha)$. Figure 6.3 presents the flowchart of adapting metaheuristics for feature selection problem.

6.3 EXPERIMENT SETTINGS

This chapter investigates the efficacy of seven well-known metaheuristic algorithms, referred to as MPA, WOA, DE, GWO, TLBO, EO, and SMA, to select the near-optimal subset of features from medical datasets which will improve both classification accuracy and computational cost. Eleven medical datasets were obtained from the UCI repository and are described in Table 6.2. These datasets were used in our

TABLE 6.2
List of the employed datasets in our experiments

ID	Dataset	N.o.F	N.o.S	N.o.C	ID	Dataset	N.o.F	N.o.S	N.o.C
1	Spect	22	267	2	7	Diabetes	8	768	2
2	Sonar	60	208	2	8	DNA	57	318	2
3	Exactly	13	1000	2	9	Wine	15	178	3
4	Exactly2	13	1000	2	10	Liver disorders	6	345	2
5	heart-statlog	13	270	2	11	Breast cancer	9	699	4
6	Blood_Transfusion	4	748	2					

experiments to evaluate the effectiveness of the algorithms in terms of three perform-ance metrics: the length of the selected feature, the classification accuracy, and the fitness value. The results of these evaluations were subjected to various statistical analyses, including the best, average, worst, and standard deviation (SD) values. On a device that has the following features: 32 gigabytes of random access memory, an Intel Core i7 processor running at 2.40 gigahertz, and a copy of Windows 10 pre-installed, each of the algorithms are implemented using the MATLAB platform with the same population size, the maximum number of function evaluations being either 40 or 100.

6.4 PERFORMANCE METRICS

The performance of various metaheuristic algorithms investigated in this chapter for the feature selection problem of medical instances to help doctors in predicting diseases accurately and carefully is evaluated using four following metrics within 20 independent runs:

    a. Classification accuracy using KNN.
    b. Fitness values.
    c. The selected number of features.
    d. Standard deviation.

6.4.1 CLASSIFICATION ACCURACY USING KNN

After the optimization process has been completed, the acquired subset of characteristics is next analysed using the K-nearest neighbour technique to determine the degree to which classification accuracy is improved by using this subset. After the calculation of the classification accuracy over 20 runs utilizing KNN, the average (Avg) of classification values across those runs is then produced. This number is then utilized to validate and compare the performance of the various methods. The ideal algorithm is the one that achieves the highest degree of accuracy.

6.4.2 FITNESS VALUES

Equation (6.2) directs the optimization process to the optimal solution, where it is used to measure the fitness of each solution; the solution that has the lowest fitness is considered to be the best, and it is used within the optimization process to direct the other solutions to the region which may contain the optimal solutions. The algorithms are each subjected to 20 separate runs, and the fitness values that are obtained after each run are totalled up and divided by the total number of runs to obtain an overall average to determine which observed metaheuristic algorithm could reach a better average.

6.4.3 THE SELECTED NUMBER OF FEATURES

The effectiveness of the algorithms is evaluated based on the number of selected features; however, this is not the primary metric being considered because a low number of features can still have low accuracy. As a consequence of this, the purpose of the process of addressing the FS problem is not only to reduce the number of features chosen but also to increase the accuracy of the classification.

6.4.4 STANDARD DEVIATION (SD)

The standard deviation (SD) is applied to the generated fitness values, classification accuracy, and selected feature to determine whether or not they have converged to measure the algorithms' consistency within 20 separate runs. The most stable algorithm is the one that has a standard deviation value that is the lowest. The standard deviation can be computed using the following formula:

$$SD = \sqrt{\frac{1}{nr-1}\sum_{i=1}^{nr}\left(f_i - \overline{f}\right)^2} \qquad (6.3)$$

where nr represents the number of runs; f_i indicates the fitness value of the ith run; and $\overline{f}$ reflects the average fitness values acquired from 20 separate runs.

6.5 CHOICE OF PARAMETERS

Regarding the other parameters of the MHAs, this chapter presents the results of numerous tests designed to identify the parameter value that is most appropriate for each method in terms of addressing the issue of feature selection. Extensive experiments under various values for each parameter have been done to achieve this goal for MPA, which has two primary parameters: P and FADs, which need to be accurately estimated to maximize its performance. The average fitness values that have been obtained under each of the various values for each parameter are depicted in Figure 6.4. Taking a closer look at this figure reveals that the optimal value for both P and FADs is 0.9 and 0.7, respectively. The performance of EO can be improved

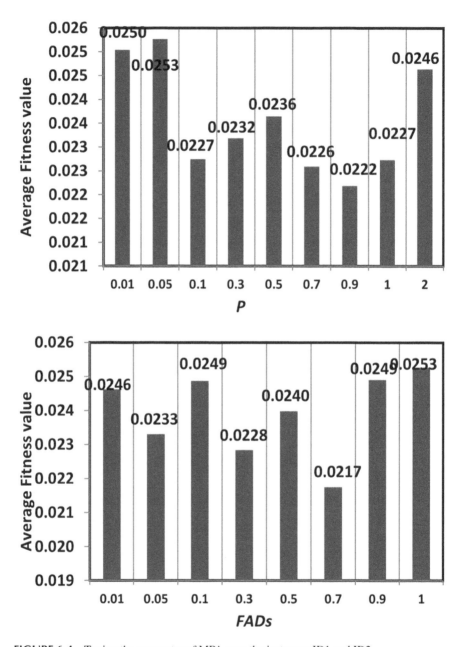

FIGURE 6.4   Tuning the parameter of MPA over the instances ID1 and ID2.

by properly predicting its four key effective parameters, denoted by the letters a_1, a_2, V, and GP, respectively. This will allow you to get the most out of the method. Consequently, a wide range of potential values for these characteristics is being investigated to identify the one that contributes most to the enhancement of EO's overall performance (see Figure 6.5). As a result of this chapter, we may conclude that the optimal values for these parameters are, in order, 5, 3, 4, and 0.4. When coming to the parameters of DE, which are F and Cr, the optimal values for them are estimated by running DE under a variety of different values for each parameter, and the results

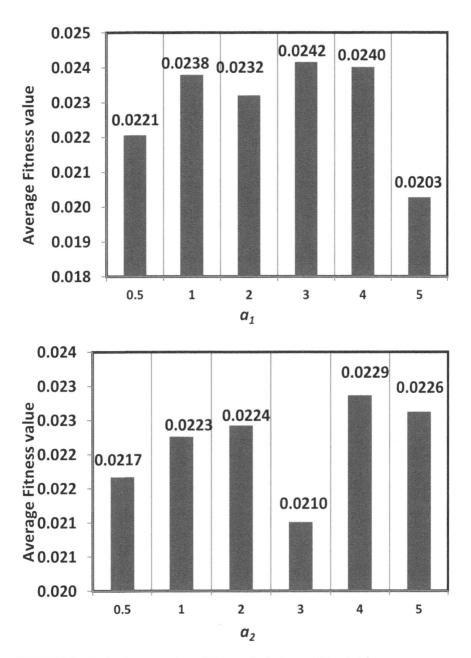

FIGURE 6.5   Tuning the parameters of EO over the instances ID1 and ID2.

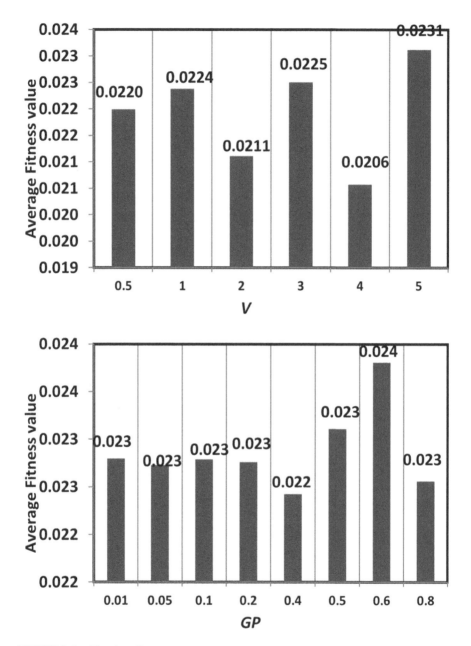

FIGURE 6.5   (Continued)

that were achieved (the average fitness value for each value of each parameter) are displayed in Figure 6.6. Based on the results shown in this figure, we determined that 1 and 0.8 were the best possible values for F and CR, respectively. Last but not least, WOA has a parameter called b and is responsible for defining the shape of the

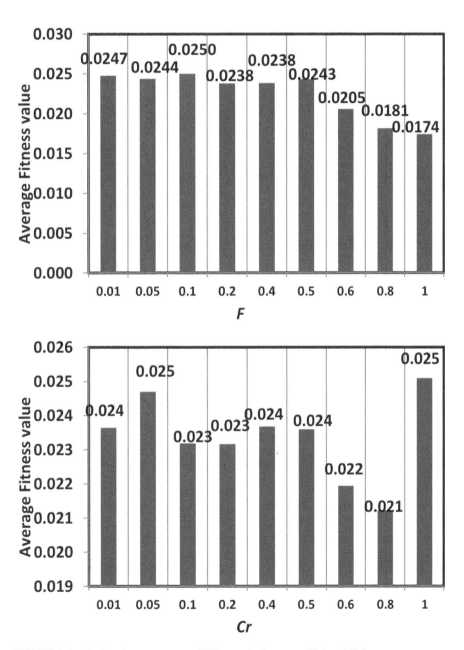

FIGURE 6.6 Tuning the parameters of DE over the instances ID1 and ID2.

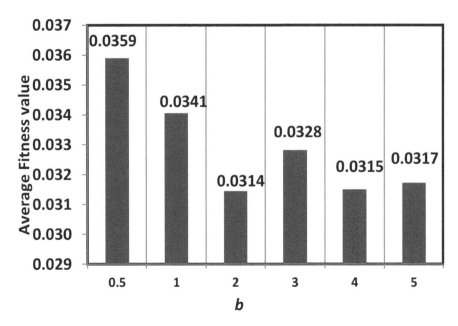

FIGURE 6.7   Tuning the WOA parameter b over ID1 and ID2.

logarithmic spiral. Tests have been conducted under various values, such as 0.5, 1, 2, 3, 4, and 5, to determine the optimal value for this parameter. The results of these experiments demonstrate that the optimal value for this parameter is 2, as shown in Figure 6.7. Finally, the single effective parameter of the SMA is indicated by the letter z. The efficiency of this parameter is evaluated using the data shown in Figure 6.8, which was produced from a large number of tests in which this parameter was set to various values. 0.05 appears to be the ideal value for this statistic, based on the data presented here.

6.6   RESULTS AND DISCUSSION

6.6.1   COMPARISON OF VARIOUS TRANSFER FUNCTIONS

This section compares the fitness values obtained by integrating each transfer function with various investigated metaheuristic algorithms to determine which transfer function is better for each algorithm. All algorithms for each transfer function were executed 20 times independently on each instance between ID#1 and ID#11, and the average fitness values for all of these instances are shown in Figure 6.9, which indicates that MPA, EO, WOA, GWO, DE, SMA, and TLBO performs better over S1, V1, V4, S3, V4, S2, and V2, respectively. At the end of this section, it is concluded that each algorithm could perform better under a specific transfer function which must be carefully determined before conducting the experiments to maximize its performance.

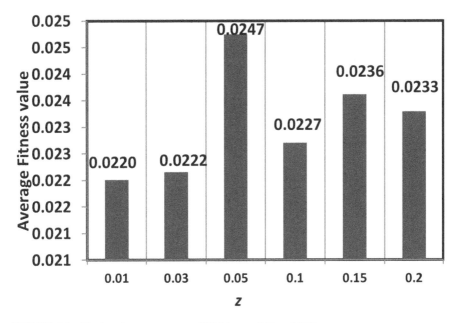

FIGURE 6.8   Tuning the parameter z of SMA over ID1 and ID2.

6.6.2   COMPARISON AMONG ALGORITHMS USING FITNESS VALUES

All algorithms are compared based on their fitness values obtained within 25 independent runs on each dataset. Those fitness values have been analysed in terms of the average, and SD values which are introduced in Figures 6.10 and 6.11. From observing those figures, it is clear that EOV1 could perform better than all the other algorithms. In addition, to show the stability of each algorithm within the independent runs on all instances, Figure 6.11 is presented to display the average of the SD obtained by each algorithm on all instances. From this figure, TLBOV2 could rank the first with a value of 0.0231, followed by both DEV4 and EOV1, while WOAV4 is a less stable one with a value of 0.038. From this analysis, it could be concluded that EOA1 could overcome the other algorithms in terms of the fitness value, which is composed of two objectives: accuracy and the number of selected features. Since the machine learning techniques rely primarily on accuracy as the first objective and the number of selected features as the second objective, the accuracy produced by various algorithms needs to be compared to determine whether EOV1 is capable of coming true with superior accuracy.

6.6.3   COMPARISON USING CLASSIFICATION ACCURACY

Figure 6.12 presents an analysis of the classification accuracy values that were achieved by each algorithm after a total of 20 independent runs on each case. The examination of this figure reveals that EOV1 could be superior to all of the competing algorithms with a classification accuracy rate up to 0.8227, followed by DEV2 with

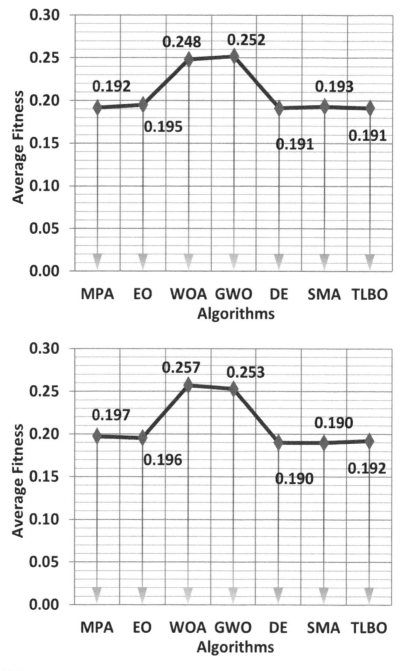

FIGURE 6.9 Average fitness values obtained by various algorithms under all observed transfer functions.

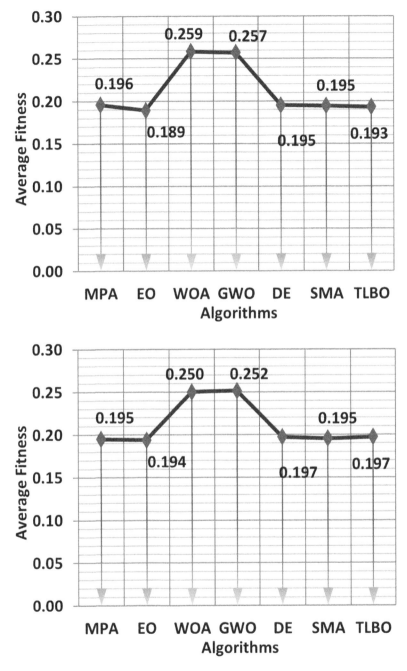

FIGURE 6.9   (Continued)

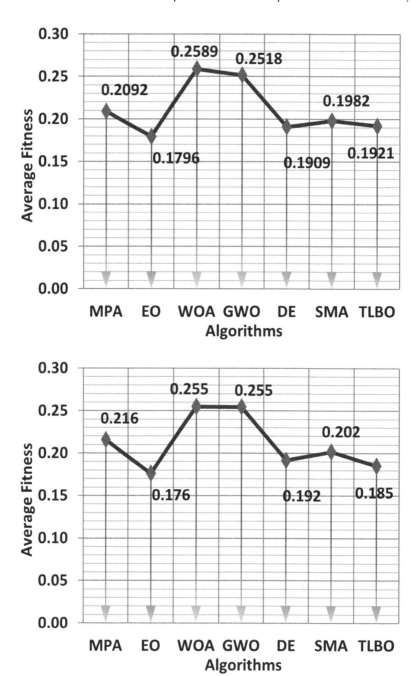

FIGURE 6.9   (Continued)

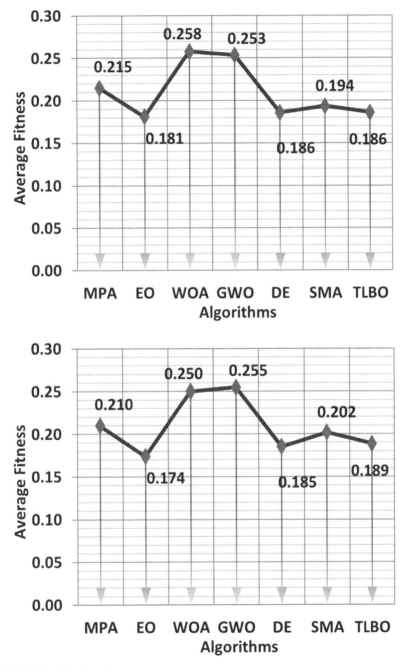

FIGURE 6.9   (Continued)

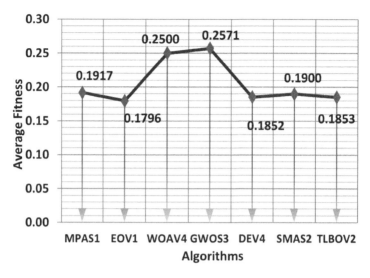

FIGURE 6.10 Comparison in terms of average fitness.

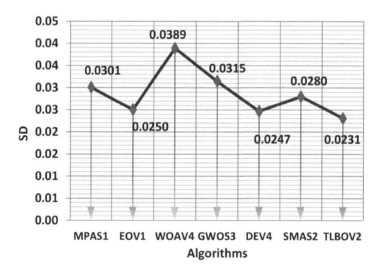

FIGURE 6.11 Comparison in terms of SD.

a value of 0.8179, demonstrating the capability of EOV1 in reaching the subset of features that maximizes the classification accuracy. In addition, the average standard deviation values for the classification accuracy under various observed datasets are presented in Figure 6.13. This figure illustrates that TLBOV4 is the more stable one, with a value up to 0.0237, whereas WOAV4 is the less stable one. Based on the findings of the earlier experiments, EOA1 has a good chance of being the optimal

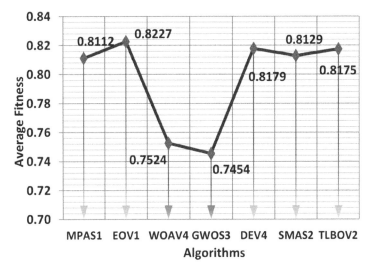

FIGURE 6.12 Comparison in terms of average accuracy rate.

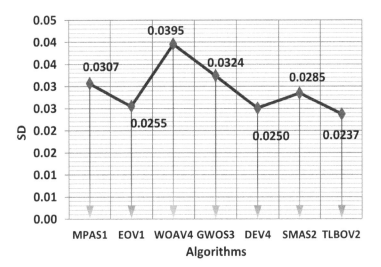

FIGURE 6.13 Comparison in terms of SD on accuracy.

solution in terms of the fitness values, which comprise the classification accuracy and the length of the features.

Furthermore, EOA1 has a good chance of being the optimal solution regarding classification accuracy, which is the primary aim of machine learning strategies. However, the length of the features is also significant since it plays a role in helping to minimize the computational cost that is consumed throughout the process of learning

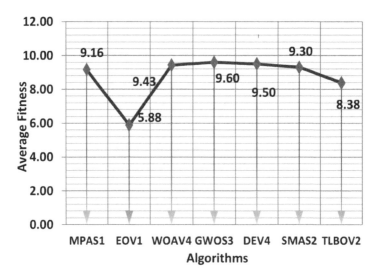

FIGURE 6.14   Comparison in terms of average selected feature.

and testing. As a result, the next part will be devoted to contrasting the algorithms with respect to the length of the selected features.

6.6.4   COMPARISON AMONG ALGORITHMS USING THE SELECTED FEATURE-LENGTH

Figures 6.14 and 6.15 are provided to demonstrate the average number of selected features and their standard deviation across all datasets. This is done so that it can be determined which algorithm has the lowest number of the selected features and which one has the largest number of the selected features overall. Based on these numbers presented in Figures 6.14 and 6.15, it can be seen that EOV1 has the potential to come in the first rank with a value of 5.88, and in fourth place after MPAS1, DEV4, and SMAS2 in terms of the average standard deviation. EOV1 is superior to all algorithms in terms of fitness value, classification accuracy, and the selected number of features. Based on that, EOV1 is a strong feature selection technique compared to all the other rival metaheuristic algorithms.

6.7   CHAPTER SUMMARY

FS is an essential step in the preprocessing of medical datasets, as it helps remove irrelevant, noisy, and redundant features while reducing the data's dimensionality. This makes it possible for machine learning algorithms to achieve higher accuracy with significantly less time spent on training. This chapter investigates the performance of seven metaheuristic optimization algorithms: WOA, GWO, MPA, DE, TLBO, SMA, and EO under various transfer functions, belonging to V-shaped and S-shaped families. These algorithms are validated on 11 datasets from the UCI repository and compared under various performance metrics to clarify which one is

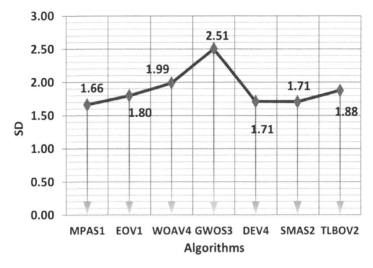

FIGURE 6.15   Comparison in terms of SD on selected feature.

strong for overcoming the feature selection problem for the medical datasets. First, the experiments have been conducted to find the relevant transfer function for each investigated metaheuristic algorithm. After that, all algorithms were compared using various performance metrics: SD, fitness values, classification accuracy, and length of selected feature to show the efficiency of each one. The experimental findings proved that EO under the transfer function V1 could reach the near-optimal subset of a feature that could come true with better classification accuracy compared to all the other algorithms. As a result, EOV1 can analyse medical data more efficiently and accurately. Thereby it could aid doctors in identifying patients more promptly and at a lower cost.

6.8   EXERCISES

a.   Explain the dimensionality reduction technique. How can that help in the feature selection problem for healthcare data?

b.   What are the common feature extraction approaches? Provide an explanation for each of them.

c.   How can a metaheuristic algorithm influence the performance of feature selection and extraction for healthcare data?

d.   Why do we need transfer functions to solve feature extraction problems? What are the common types of transfer functions?

REFERENCES

1.   Sagban, R., et al., Hybrid bat-ant colony optimization algorithm for rule-based feature selection in health care. *International Journal of Electrical and Computer Engineering*, 2020. **10**(6): pp. 6655–6663.

2. Hall, M.A. and L.A. Smith. Feature selection for machine learning: comparing a correlation-based filter approach to the wrapper. In *FLAIRS Conference*. 1999.

3. Cai, J., et al., Feature selection in machine learning: A new perspective. *Neurocomputing*, 2018. **300**: pp. 70–79.

4. Mafarja, M., et al., Efficient hybrid nature-inspired binary optimizers for feature selection. *Cognitive Computation,* 2020. **12**(1): pp. 150–175.

5. Abd Elminaam, D.S., et al., An efficient marine predators algorithm for feature selection. 2021. **9**: pp. 60136–60153.

6. Kotsiantis, S.B., I. Zaharakis, and P.J. Pintelas, Supervised machine learning: A review of classification techniques. *Emerging Artificial Intelligence Applications in Computer Engineering,* 2007. **160**(1): p. 3–24.

7. Barlow, H.B., Unsupervised learning. *Neural Computation,* 1989. **1**(3): pp. 295–311.

8. Zhu, X., A.B. Goldberg, Introduction to semi-supervised learning. 2009. **3**(1): pp. 1–130.

9. Gadat, S. and L. Younes, A stochastic algorithm for feature selection in pattern recognition. *Journal of Machine Learning Research,* 2007. **8**(Mar): p. 509–547.

10. Arumugam, K., S. Ramasamy, and D. Subramani, Binary duck travel optimization algorithm for feature selection in breast cancer dataset problem. In *IOT with Smart Systems*. 2022, Springer. pp. 157–167.

11. Sayed, G.I., A.E. Hassanien, and A.T. Azar, Feature selection via a novel chaotic crow search algorithm. *Neural Computing and Applications*, 2019. **31**(1): p. 171–188.

12. Ghosh, K.K., et al., Improved binary sailfish optimizer based on adaptive β-hill climbing for feature selection. *IEEE Access*, 2020. **8**: p. 83548–83560.

13. Anand, P. and S. Arora, A novel chaotic selfish herd optimizer for global optimization and feature selection. *Artificial Intelligence Review,* 2020. **53**(2): p. 1441–1486.

14. Manikandan, R. and A. Kalpana, Feature selection using fish swarm optimization in big data. *Cluster Computing,* 2019. **22**(5): pp. 10825–10837.

15. Sayed, G.I., A. Tharwat, and A.E. Hassanien, Chaotic dragonfly algorithm: an improved metaheuristic algorithm for feature selection. *Applied Intelligence,* 2019. **49**(1): pp. 188–205.

16. Hussien, A.G., et al., S-shaped binary whale optimization algorithm for feature selection, in *Recent Trends in Signal and Image Processing*. 2019, Springer. pp. 79–87.

17. Mafarja, M., et al. Feature selection using binary particle swarm optimization with time varying inertia weight strategies. In *Proceedings of the 2nd International Conference on Future Networks and Distributed Systems*. 2018.

18. Abdel-Basset, M., et al., A new fusion of grey wolf optimizer algorithm with a two-phase mutation for feature selection. *Expert Systems with Applications,* 2020. **139**: p. 112824.

19. Kennedy, J. and R.C. Eberhart. A discrete binary version of the particle swarm algorithm. in *1997 IEEE International Conference on Systems, Man, and Cybernetics. Computational Cybernetics and Simulation*. 1997. IEEE.

20. Yadav, S. and S. Shukla. Analysis of k-fold cross-validation over hold-out validation on colossal datasets for quality classification. In *2016 IEEE 6th International Conference on Advanced Computing (IACC)*. 2016. IEEE.

21. Peterson, L.E., K-nearest neighbor. *Scholarpedia*, 2009. **4**(2): p. 1883.

7 Advanced Metaheuristics for Task Scheduling in Healthcare IoT

7.1 TASK SCHEDULING IN HEALTHCARE IOT

In the future, everything will be connected to the Internet, which will be made possible by a technology known as the Internet of Things (IoT), which is the future of the Internet and one of the essential technologies employed in recent times in all domains. One of its applications is the use of IoT technology in healthcare, specifically in remote patient monitoring and mobile health for patients suffering from various diseases, including cardiovascular disease, high blood pressure, diabetes, and other chronic conditions. It is undesirable for any delay caused by sending data to and from the cloud when dealing with remote health monitoring applications since the real-time aspect of these applications is of the utmost importance. Because healthcare applications are sensitive to latency, display poor response time, and output a big amount of data, fog computing is believed to be the ideal way to rely on when creating these applications. Providing elderly patients with home nursing care is one of the key contributions that fog computing makes to medical applications [1]. Fog computing has been utilized between sensors and cloud computing to gather and effectively process data. This will reduce the quantity of data that is carried between the cloud and the sensors, and it will also boost the efficiency of the entire system. Wireless sensor networks (WSNs), which are utilized in the health monitoring field, simultaneously send many jobs to fog computing, each of which varies in terms of its importance and length. As a result, we need to put in place an adequate task scheduling algorithm that can accurately schedule the tasks in a form that will reduce the response time to the patients because this application is sensitive to latency.

Therefore, to solve this issue, the researchers who have contributed to the body of published work have utilized a variety of conventional, heuristic, and metaheuristic optimization strategies. However, to overcome the computational challenge, the only hope lies in using heuristic and metaheuristic algorithms, which have been utilized in the past to tackle multiple complex and large non-deterministic polynomial (NP) problems to achieve significant results in a reasonable amount of time [2, 3]. This is because this problem is classified as NP-hard, which makes it extremely challenging to solve using traditional methods. Several heuristics and metaheuristic algorithms have been proposed for tackling this problem, some of which are the Harris hawks optimization algorithm (HHO) [4], ant colony optimization (ACO) [5], fireworks

DOI: 10.1201/9781003325574-7

algorithm (FWA) [6], non-dominated sorting genetic algorithm (NSGA-II) [7], bees algorithm (BA) [7], hybrid metaheuristic algorithm [8], marine predators algorithm (MPA) [9], and improved elitism-based genetic algorithm (IEGA) [10].

This chapter investigates the performance of six metaheuristic optimization algorithms for finding the near-optimal scheduling of tasks to the virtual machines (VMs) in fog computing to improve the quality of service presented to the patients in the healthcare system. Various performance metrics: make-span, energy consumption, flow-time, and carbon dioxide emission rate are considered in this chapter to observe the performance of each metaheuristic algorithm. This chapter is organized as follows: at the beginning of this chapter, the problem formulation will be presented, in addition to a description of how to alter the metaheuristic algorithms to solve this problem typically. Controlling parameters and the experimental settings for the designed task scheduling problem are given at the later stage. In the end, but certainly not least, a few experiments are carried out to determine which algorithm would be the most effective solution. At long last, a summary of the chapter is offered.

7.2   PROBLEM FORMULATION

The task scheduling problem in fog computing is considered a multi-objective problem since it includes more than one objective that must be optimized simultaneously. These objectives which are herein named as performance metrics, are make-span, flow-time, energy consumption, and carbon dioxide emission rate. Each of which will be discussed in detail within the next sections. In case of the task scheduling problem in fog computing, assume that there are d different tasks that have been sent to the fog computing system, which has about M different fog devices. The tasks need to be assigned to those M different fog devices in an order that will minimize the performance metrics that have been mentioned previously to improve the overall quality of the services.

7.2.1 Make-Span

Before beginning the task scheduling, a starting time of 0 is allotted to each fog device in a fog computing system. Subsequently, the tasks that need to be processed are distributed among the VMs in the system for them to be completed. After all of the tasks that were assigned to each machine have been finished, the amount of time that each machine spent in execution will be calculated, and the machine that took the most time in execution Et_j will be the one that is used to signify the make-span. The make-span is the maximum amount of execution time that a VM uses to complete the tasks that have been given to it.

7.2.2 Energy Consumption

There are two states for each virtual machine: active and inactive or idle. This VM's energy consumption in the idle state will reach 60% of its energy consumption in the active state. To calculate the energy consumption of each VM, both active and idle states must be taken into account, as shown in the following formula:

$$E\left(VM_j\right) = \left(Et_j \times b_j + \left(MK - Et_j\right) * a_j\right) \times s_j \qquad (7.1)$$

$$b_j = 10^{-8} \times \left(s_j^2\right) \qquad (7.2)$$

$$a_j = 0.6 * b_j \qquad (7.3)$$

Where b_j indicates the amount of energy consumed in the active state and s_j stands for processing speed of the *jth* VM measured by million instructions per second (MIPS) [11].

7.2.3 CARBON DIOXIDE EMISSION RATE (CDER)

The following formula is used to calculate the amount of carbon dioxide emission rate emitted by each VM to determine if it is environmentally friendly:

$$carbon\_emission\_rate = \sum_{k=1}^{4} total\_energy \times sh_k \times emission\_factor_k \times ratio \qquad (7.4)$$

The power comes from a variety of sources, including coal, natural gas, oil, and non-fossil products, which are designated by the numbers $k = 1, 2, 3$, and 4, respectively. The percentage of the total amount of energy that comes from the kth the different source is denoted by the symbol sh_k. In addition, the term $emission\_factor_k$ is used to express the emission factor of the kth energy source. The value of this term is 0.5825 for oil, 0.7476 for coal, 0.4435 for natural gas, and 0 for non-fossil products [12]. The ratio equals 44/12, and this number represents the ratio for turning carbon into carbon dioxide [13].

7.2.4 FLOW-TIME (FT)

This statistic calculates the overall execution time required by all VMs to complete their assigned tasks. This metric is computed according to the following formula:

$$FT = \sum_{i=0}^{n-1} f_i \qquad (7.5)$$

where f_i indicates the overall execution time required by the ith VM to complete its tasks.

7.3 ADAPTATION OF METAHEURISTICS FOR THE TASK SCHEDULING PROBLEM

At the beginning of the optimization procedure, each investigated metaheuristic algorithm must distribute its solutions within the problem's bounds. When applying the

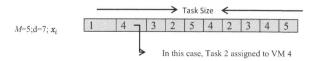

FIGURE 7.1   Solution representation to TSFC problem.

optimization technique to combinatorial issues, such as the task scheduling problem in fog computing (TSFC), a normalization approach is required to turn continuous values into discrete values that reflect the range of the problem. The following describes how to initialize the solutions within the problem's upper and lower bounds. In the discrete TSFC problem, a group consisting of N solutions is provided during the initialization step of the optimization process. In this group, each solution x_i is compounded of d dimensions. Each solution x_i in the group provides an individual solution to the TSFC problem, where the d dimensions in each solution specify the number of jobs that need to be allocated to the VMs. During the process of initializing each solution, assuming that there are M virtual machines, each job will be arbitrarily assigned to one of the M VMs. Figure 7.1 presents an example to show how to represent a solution for the TSFC problem.

After completing the initialization process, the quality of each solution must be calculated to identify the best-so-far solution which could minimize the fitness function. This function is based on both make-span and energy consumption as the two factors affecting the carbon dioxide rate and flow time.

$$f\left(\overrightarrow{X_i}\right) = \tau \times \text{total\_energy} + \left(1 - \tau\right) \times \text{MK} \qquad (7.6)$$

where τ is employed to trade off between the energy consumed and make-span as the two effective objectives for tackling this problem. In this chapter, we employ a value of 0.2 for this parameter.

The optimization process of each metaheuristic algorithm will then be executed to update the initialized solutions to get a better solution. However, the produced solutions by the algorithms are continuous, which contradicts the combinatorial nature of the TSFC issue. Consequently, Equation (7.7) will be used to normalize these continuous solutions between 0 and 1. The normalized solutions are then scaled between 1 and M (the number of virtual machines) using Equation (7.8).

$$\vec{X}'_i\left(t+1\right) = \frac{X_i\left(t+1\right) - L}{U - L} \qquad (7.7)$$

$$\vec{X}''_i\left(t+1\right) = \vec{X}'_i\left(t+1\right) * M \qquad (7.8)$$

L and U represent the minimum and maximum values in the modified solution $\vec{X}_i\left(t+1\right)$. The normalized solution $\vec{X}''_i\left(t+1\right)$ will then be evaluated using the prior performance criteria and compared to the other solutions to determine which is superior.

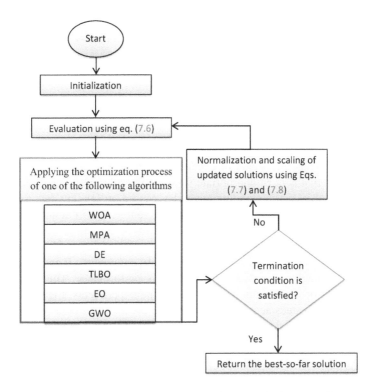

FIGURE 7.2   Flowchart of adapting metaheuristics for task scheduling problem.

7.4   EXPERIMENTAL SETTINGS

The efficiency of seven different metaheuristic algorithms that have been detailed in the past, including WOA, GWO, SMA, EO, DE, TLBO, and MPA, has been evaluated through the course of several different tests that have been conducted using two various benchmarks. The first benchmark consists of 12 different tasks with the following sizes: 100, 200, 300, 400, 500, 600, 700, 800, 900, 1000, 1200, and 1500, while the total number of VMs that are accessible remains constant and is always equal to 100. The amount of work that needs to be done for each task is unpredictable and might range anywhere from 1000 to 10,000. The processing speed of the VMs is split in half, with each half receiving a different amount of MIPS: the first half receives 2000 MIPS, while the other half receives 4000 MIPS. In addition, the price of each VM is proportionally related to the processing speeds. Depending on the processing speed of each VM, the first half of the price will be set to 200 dollars per time unit, while the other will be 400 dollars. The second benchmark consists of a task with a fixed size of 600, and the lengths of the virtual machine that will be used to complete this task will be 10, 20, 30, 40, 50, 60, 70, 80, 90, 100, 120, and 150 respectively. The processing speed of the VMs may be broken down into two halves: the first has a processing speed of 2000 MIPS and costs 200 dollars, while the second has a speed of 4000 MIPS and costs 400 dollars.

All of the experiments are carried out using a computer equipped with 32 gigabytes of random access memory (RAM), a 2.40 GHz Intel Core i7-4700MQ CPU, and a Windows 10 operating system. All of the algorithms investigated in this chapter are implemented using Java programming language, with a population size of 30 and a maximum number of iterations of 1500. In addition to this, statistical analyses are utilized to validate the obtained findings.

7.4.1 CHOICE OF PARAMETERS

This chapter presents the results of extensive experiments designed to identify the parameter value that is most appropriate for each algorithm in terms of its ability to solve the TSFC to assist in avoiding any delay caused by sending data to and from the cloud when dealing with applications that do remote health monitoring because the real-time aspect of these applications is of the utmost importance. Extensive experiments under various values for each parameter have been conducted to achieve this objective for the MPA, which has two primary parameters: P and FADs, which need to be accurately estimated to maximize its performance. This objective has been the subject of much attention throughout the experiments. Figure 7.3 depicts the average fitness values acquired under each of the many different values for each parameter. When you take a closer look at this figure, you'll notice that the ideal value for both P is 1, while the ideal value for FADs is 0.80. To get the most out of the EO technique, its performance can be improved by properly predicting its four key effective parameters, which are denoted by the letters a_1, a_2, V, and GP, respectively. This will allow you to get the most out of the method. Consequently, a wide variety of possible values for these parameters are now being investigated to identify the one that contributes the most to an improvement in the overall performance of EO (see Figure 7.4). As a result of looking at this figure, we may conclude that the optimal values for these parameters are, in order, 1.5, 4, 3, and 0.2. This would be the case because these values produce the best results. When it comes to the parameters of DE, which are F and Cr, the optimal values for them are estimated by running DE under a range of different values for each parameter. The outcomes that were attained (the average fitness value for each value of each parameter) are shown in Figure 7.5. Following an analysis of the data presented in this figure, we concluded that the optimal values for F and CR, respectively, were 0.01 and 0.05. In conclusion, but certainly not least, the WOA only uses one useful parameter, denoted by b. This parameter is responsible for establishing the shape of the logarithmic spiral. Several experiments have been carried out using different values for this parameter, such as 0.5, 1, 2, 3, 4, and 5, to identify which of these numbers is the most appropriate. As can be seen in Figure 7.6, the outcomes of these experiments indicate that the best possible value for this parameter is 1.

In conclusion, the SMA only uses a single effective parameter, represented by the letter z. The effectiveness of this parameter is evaluated using the data shown in Figure 7.7, which was produced from a large number of experiments in which different values were used for this parameter. These tests were carried out to examine

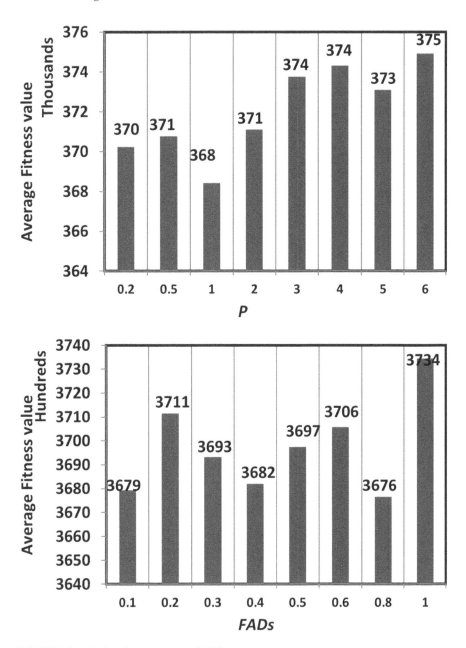

FIGURE 7.3   Tuning the parameter of MPA.

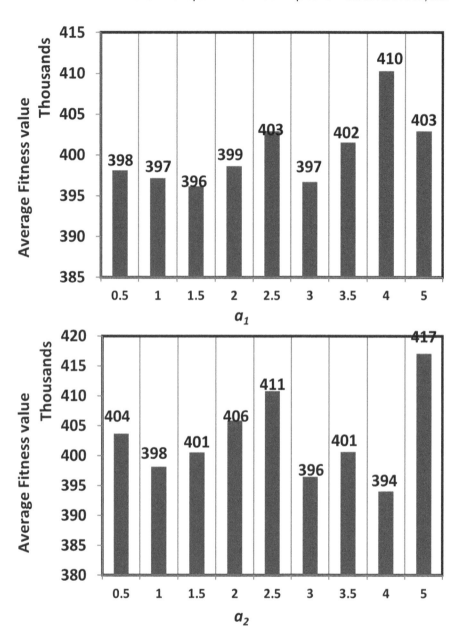

FIGURE 7.4   Tuning the parameters of EO.

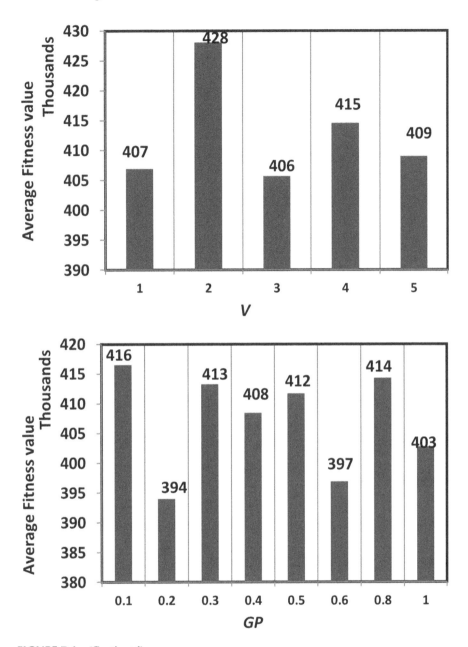

FIGURE 7.4 (Continued)

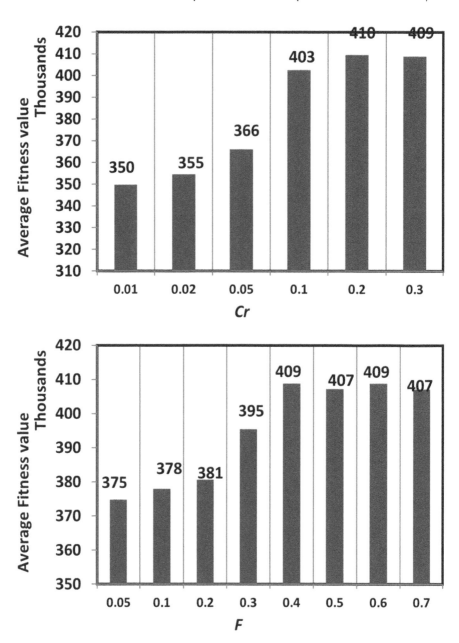

FIGURE 7.5   Tuning the parameter of DE.

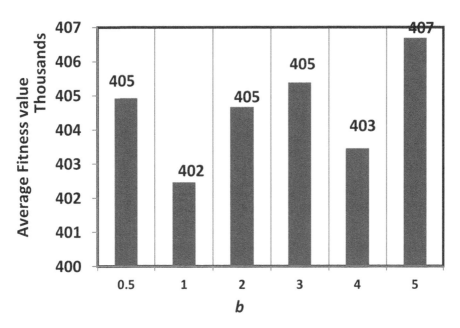

FIGURE 7.6   Tuning the parameter b of WOA.

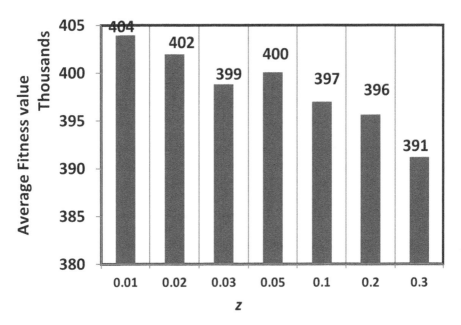

FIGURE 7.7   Tuning the parameter z of SMA.

the effects of varying this parameter's value. According to the data presented in this figure, the parameter in question would benefit most from having the value of 0.3 assigned to it.

7.5 RESULTS AND DISCUSSION

In this section, all algorithms will be compared using three performance metrics, namely make-span, energy consumption, and carbon dioxide commission rate, to determine each one's effectiveness in tackling the task scheduling problem in fog computing.

7.5.1 COMPARISON UNDER VARIOUS TASK SIZES

In this section, the performance of seven various metaheuristic algorithms is evaluated using the first benchmark, which contains a variety of task sizes (TS) but has the same virtual machine length (VL). This evaluation is done to identify how well each investigated algorithm performs. Each algorithm that was utilized in our experiments was run a total of 20 times. Subsequently, the average fitness value (F-value) and the carbon dioxide emission rate (CDER), make-span, and energy for each of those 30 runs were calculated and presented in Figure 7.8. Figure 7.8(a) displays the average fitness value obtained by each algorithm within 20 independent times to show readers which algorithm performs better under various task sizes: small scale, medium scale, and large scale. According to this figure, MPA has the potential to achieve the first rank with a total of 762567, while GWO achieves the position of last place with a value of 906572. As further evidence, we present Figure 7.8(b), which illustrates an average of CDER emitted based on the task scheduling obtained by each algorithm. This figure demonstrates that MPA is capable of achieving better task scheduling, which could distribute the tasks to the fog nodes in a manner that lowers the rate of carbon dioxide emitted, thereby demonstrating that MPA is more environmentally friendly.

Make-span is a well-known statistic that is frequently applied to scheduling challenges. This metric measures the execution time required by the final VM to complete all assigned tasks. It can be decreased by scheduling IoT tasks to improve patient service quality and reduce potential delays in healthcare systems. Consequently, all algorithms were evaluated using this metric to see which algorithm could schedule the activities to minimize the make-span. Regarding the make-span measure provided in Figure 7.8(c), this figure reveals that MPA is superior to the other algorithms, with a value of 27.01 compared to the other algorithms' values. Energy is one of the essential aspects of our daily lives, and no one on earth could survive without it. Therefore, conserving energy is crucial for the continuation of our daily lives. Consequently, the average amount of energy consumed by each algorithm during 20 independent runs on all instances is shown in Figure 7.8(d), demonstrating that MPA could come in first place after defeating all other algorithms with a score of 759618.

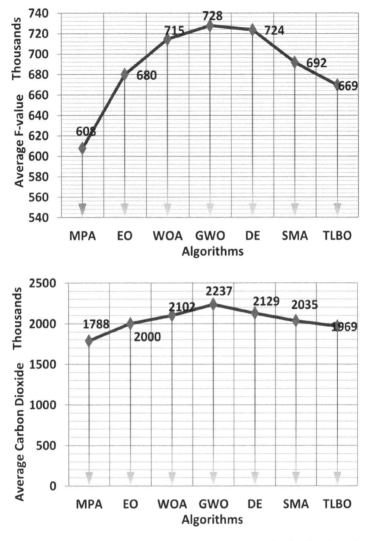

FIGURE 7.8   Comparison of various performance metrics over the first benchmark.

7.5.2   COMPARISON UNDER VARIOUS VM LENGTHS

In this part, the performance of seven distinct metaheuristic algorithms is tested utilizing the second benchmark, which includes VMs of varying lengths but the same task size. This examination is conducted to determine the performance of each algorithm under investigation. Each algorithm employed in our tests was executed 30 times. Consequently, the average fitness value (F-value), carbon dioxide emission rate (CDER), make-span, and energy for each of the 20 runs were determined and

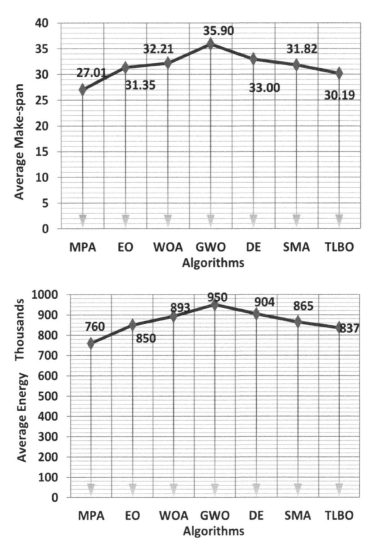

FIGURE 7.8   (Continued)

displayed in Figure 7.9. In a general sense, Figure 7.9(a) depicts the average fitness value obtained by each algorithm over 20 independent trials to demonstrate which algorithm has superior performance for small-scale, medium-scale, and large-scale tasks. According to this number, MPA can attain first place with a score of 438,500, while GWO lands in last place with a score of 529,483. As additional evidence, we give Figure 7.9(b), which depicts the average amount of CDER emitted for each algorithm based on the task scheduling determined. This figure indicates that MPA is capable of improved work scheduling, which could allocate tasks to fog nodes in

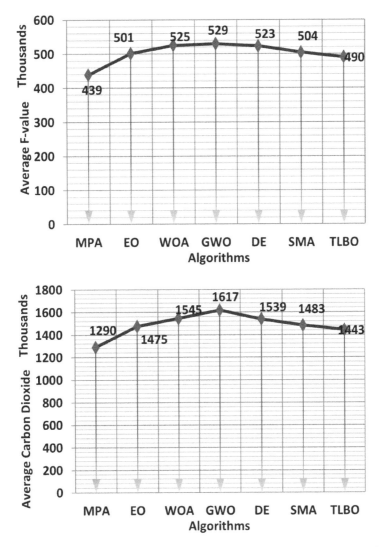

FIGURE 7.9 Comparison of various performance metrics over the second benchmark.

a manner that reduces the amount of carbon dioxide emitted, thereby indicating that MPA is more eco-friendly.

Regarding the make-span metric presented in Figure 7.9(c), this figure demonstrates that MPA is superior to the other algorithms, with a value of 43 compared to the values of the other algorithms. Energy conservation is essential for our continued existence. Consequently, the average amount of energy consumed by each algorithm throughout 20 independent runs on all instances is depicted in Figure 7.9(d), which reveals that MPA was able to position first by outperforming all other algorithms with a score of 548,114.

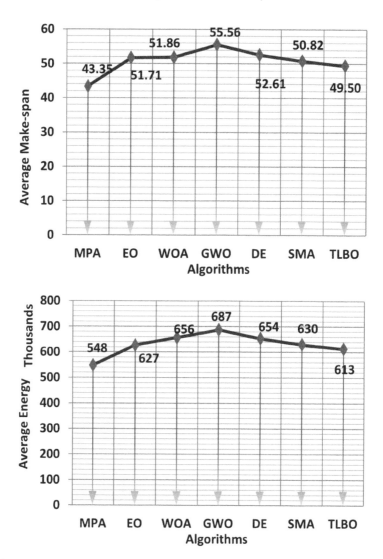

FIGURE 7.9   (Continued)

7.6   CHAPTER SUMMARY

This chapter discusses the role of metaheuristic algorithms for tackling the task scheduling problem in fog computing to see their effectiveness in reaching the near-optimal
scheduling of tasks in a form that will minimize the delay caused by processing the
tasks in the fog nodes. This will be beneficial for several applications, especially real-
time applications such as remote patient monitoring and mobile health for patients
suffering from a variety of diseases, including cardiovascular disease, high blood
pressure, diabetes, and other chronic conditions. MHAs have a strong performance for overcoming this problem, and to show that practically, in this chapter, we

selected seven well-known MHAs, which are selected based on into factors: recently published and highly cited. Those algorithms are WOA, DE, MPA, GWO, TLBO, SMA, and EO. After that, these algorithms have been evaluated through the course of several different tests that have been conducted using two various benchmarks. The algorithms were compared in terms of the average fitness value (F-value), as well as the carbon dioxide emission rate (CDER), make-span, and energy. Finally, the experimental findings show that MPA performs better than all the other algorithms. As a result, MPA is a strong alternative scheduling algorithm for distributing the tasks to the fog nodes in a form that will minimize the patients' response.

7.7 EXERCISES

a. Why a task scheduling algorithm is important and how that can help in healthcare?
b. How can a metaheuristic algorithm influence the performance of task scheduling in healthcare IoT?
c. Name a few performance targets for task scheduling algorithms. Why they are important and how they should be incorporated into the task scheduling problem?
d. How can we adapt the metaheuristic algorithm to the task scheduling problem? Provide an example
e. Why a proper tuning of algorithmic parameters is important? How can we do that?

REFERENCES

1. Mouradian, C., et al., A comprehensive survey on fog computing: State-of-the-art and research challenges. *IEEE Communications Surveys & Tutorials*, 2017. **20**(1): pp. 416–464.
2. Halim, Z., Optimizing the DNA fragment assembly using metaheuristic-based overlap layout consensus approach. *Applied Soft Computing*, 2020. **92**: p. 106256.
3. Mousavirad, S.J. and H. Ebrahimpour-Komleh, Multilevel image thresholding using entropy of histogram and recently developed population-based metaheuristic algorithms. *Evolutionary Intelligence*, 2017. **10**(1): pp. 45–75.
4. Abdel-Basset, M., et al., Energy-aware metaheuristic algorithm for industrial Internet of Things task scheduling problems in fog computing applications. *IEEE Internet of Things Journal*, 2020. (99): pp. 1–11.
5. Kishor, A. and C. Chakarbarty, Task offloading in fog computing for using smart ant colony optimization. *Wireless Personal Communications*, 2021. **127**: pp. 1683–1704.
6. Yadav, A.M., K.N. Tripathi, and S.C. Sharma, A bi-objective task scheduling approach in fog computing using hybrid fireworks algorithm. *The Journal of Supercomputing*, 2021: p. 1–25.
7. Keshavarznejad, M., M.H. Rezvani, and S. Adabi, Delay-aware optimization of energy consumption for task offloading in fog environments using metaheuristic algorithms. *Cluster Computing*, 2021: pp. 1–29.
8. Hosseinioun, P., et al., A new energy-aware tasks scheduling approach in fog computing using hybrid meta-heuristic algorithm. *Journal of Parallel and Distributed Computing*, 2020. **143**: pp. 88–96.

9. Abdel-Basset, M., et al., Energy-aware marine predators algorithm for task scheduling in IoT-based fog computing applications. *IEEE Transactions on Industrial Informatics,* 2020. **17**(7): pp. 5068–5076.

10. Abdel-Basset, M., R. Mohamed, R.K. Chakrabortty, and M.J. Ryan. IEGA: An improved elitism-based genetic algorithm for task scheduling problem in fog computing. *International Journal of Intelligent Systems*, 2021. 36: pp. 4592–4631. https://doi.org/10.1002/int.22470.

11. Abdel-Basset, M., et al., Energy-aware metaheuristic algorithm for industrial Internet of Things task scheduling problems in fog computing applications. *IEEE Internet of Things Journal*, 2020. **8**: pp. 12638–12649.

12. Zhang, Y.-J., et al., The impact of economic growth, industrial structure and urbanization on carbon emission intensity in China. *Natural Hazards,* 2014. **73**(2): pp. 579–595.

13. Yang, H.-J. and J. He. Analysis of the ethnic minority living activities CO_2 emissions from 2012 to 2014: A case study in Yunnan. In *Advanced Materials and Energy Sustainability: Proceedings of the 2016 International Conference on Advanced Materials and Energy Sustainability (AMES2016)*. 2017. World Scientific.

8 Metaheuristics for Augmenting Machine Learning Models to Process Healthcare Data

8.1 DATA MINING FOR HEALTHCARE DATA

Data mining can be utilized in the healthcare field, particularly in datasets pertaining to cancer, cardiovascular, and heart disorders, which are currently considered to be trending subjects. Various data mining approaches are used to extract the data and patterns that help the enhancement of medical decision-making [1, 2]. One of the modern ways of predicting, detecting, and making a medical decision without the involvement of humans is known as machine learning, abbreviated ML. Machine learning applications in the medical field are rapidly expanding and now include disease detection, visualization, and the investigation of disease transmission.

ML is a subfield of artificial intelligence (AI) that is primarily concerned with creating algorithms that will enable computers to learn independently from the data they are given and from their previous experiences. ML algorithms construct a mathematical model with the assistance of the training data to help in making predictions or judgments without being explicitly programmed. Computer science and statistics fields are brought together through ML to develop predictive models. ML can be broadly divided into supervised, unsupervised, and reinforcement. Supervised learning is one in which we feed the machine learning system with examples of labelled data to train it, and then the system uses this information to make predictions about the output. Figure 8.1 depicts the framework of supervised ML. In this figure, the ML algorithm is given the training data and its associated output, called the label. Then, this algorithm is trained on this training data in an attempt to reach the given label corresponding to each record in the training data. Some of the well-known supervised ML algorithms are k-nearest neighbours, random forest, logistic regression, and decision tree.

Unsupervised learning is a learning technique in which a computer is allowed to study on its own without being supervised in any way. In unsupervised learning, the ML algorithm takes the training data that has not been labelled, categorized, or classified. The algorithm seeks to reorganize the input data into new features or a set of objects with similar patterns. One of the well-known unsupervised learning algorithms is k-mean clustering [3]. Figure 8.2 depicts the framework of unsupervised machine learning.

DOI: 10.1201/9781003325574-8

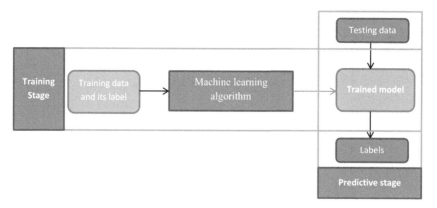

FIGURE 8.1   Framework of supervised learning.

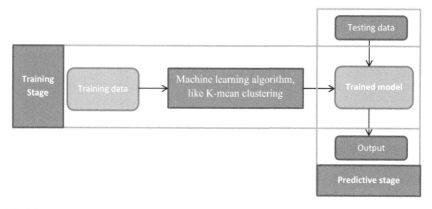

FIGURE 8.2   Framework of unsupervised learning.

Another learning technique that is based on feedback and in which a learning agent receives a reward for each correct action and a penalty for each incorrect action is referred to as reinforcement learning. The agent's performance is enhanced due to the automatic learning that occurs as a result of this feedback. During the reinforcement learning process, the agent engages in an activity with its surroundings and investigates them. An agent's purpose is to maximize the reward points it receives to raise its overall performance. The robotic dog is a good illustration of reinforcement learning because it can automatically learn how to use its arms differently.

A support vector machine (SVM) is a supervised learning system that solves problems involving classification and regression. It is predominantly applied to categorization issues [4]. The SVM algorithm aims to locate a hyperplane in an N-dimensional space that can classify the data samples. The number of features determines which dimension of the hyperplane will be used. If only two input features exist, then the hyperplane is nothing more than a straight line. If three features are used as input, then the hyperplane transforms into a two-dimensional plane. Given

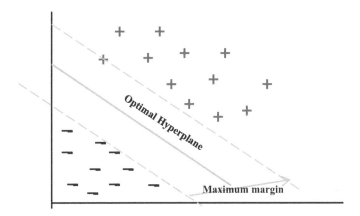

FIGURE 8.3 Illustration of two-dimensional hyperplane support vector machine.

labelled training data, the SVM generates the optimal hyperplane for classifying fresh instances. In this case, the hyperplane is a line that divides a plane into two sections, one for each class, as shown in Figure 8.3.

SVM has been utilized to address numerous pattern identification issues in various industries, including fault diagnostics and machinery condition monitoring, bioinformatics, digital image processing, handwriting recognition, time series forecasting, and financial and bankruptcy prediction [5]. SVMs have a parameter, namely regularization constant C, that defines the trade-off between reducing the training error rate as much as possible and increasing the classification margin as much as possible. In addition to this, SVMs have kernel parameters, which determine the nonlinear transition from the input feature space to a space with a higher dimension. The kernel parameters affect the performance of the SVM. Generally, the performance of SVM is controlled by selecting appropriate values for these two parameters, namely the penalty parameters and the kernel parameters [6].

Over the last decades, metaheuristic algorithms have a significant role in finding the optimal values for these parameters to fulfil high classification performance. Some of the metaheuristic algorithms applied to tackle this problem are a hybrid genetic algorithm (GA), ant colony algorithm (ACO), bat algorithm (BA), whale optimization algorithm (WOA), firefly algorithm (FFA), fruit fly optimization algorithm (FFO), particle swarm optimization (PSO), dragonfly algorithm (DFA), sine cosine algorithm (SCA) [7], differential evolution (DE), elephant herding optimization (EHO) [8], enhanced fireworks algorithm (EFWA) [9], social-spider optimization (SSO) [10], Quantum-behaved particle swarm optimization (QPSO) [11], and ant lion optimizer (CALO) [6]. There are several other works done recently for estimating these parameters; for example, in [12], the antlion optimizer has been improved to estimate these two parameters of SVM for improving the classification accuracy of lithium battery state of health. Some of the other recently published are oppositional grasshopper optimization (OGHO) [13], improved particle swarm optimization (IPSO) [14], and crow search algorithm [15]. In this chapter, we will investigate the performance of seven metaheuristic algorithms, known as WOA, MPA, EO, SMA, TLBO,

DE, and GWO for optimizing two parameters of SVM to achieve better classification accuracy when classifying medical data.

Therefore, at the beginning of this chapter, the SVM will be carefully overviewed, along with a discussion of how to apply metaheuristic algorithms to better tune the control parameters. The presentation of the choice of regulating parameters and the experimental conditions comes next after that. In conclusion, but certainly not least, a few experiments are carried out to identify which algorithm would be the most successful in terms of finding a solution to this problem. Finally, a conclusion of the chapter is provided for your perusal.

8.2   OVERVIEW OF THE SVM APPROACH

Support vector machines, abbreviated as SVMs, are one of the supervised learning techniques that are utilized the most frequently. The primary purpose of them is to differentiate between the various classes by utilizing hyperplanes. Nevertheless, the performance of SVMs is significantly impacted by nonlinearly separable data, and this issue can be remedied by utilizing kernel functions. Kernel functions are designed to map the present features onto a higher-dimensional space in which the data may be linearly separated. One of the obstacles to employing SVMs is selecting the suitable kernel function, and the other is modifying the parameters of that function. Computationally speaking, locating the optimal decision plane is an optimization issue that needs to be solved to assist the kernel functions in locating the ideal space in which the classes can be linearly divided by means of a nonlinear transformation [16].

Let's say we have M training samples $(S = S_1, S_2, S_3, \ldots \ldots, S_M)$, such that S_i represents the ith training sample, which contains d features, and it falls into one of two classes $(y_i \in \{-1, +1\})$. As a result, the training set is represented by the notation $((S_1, y_1), (S_2, y_2), (S_3, y_3), \ldots \ldots, (S_M, y_M))$, where $\{y_1, y_2, \ldots \ldots, y_M\}$ stand for the class labels associated with $S_1, S_2, \ldots$, and S_M. The decision boundary between the two classes: -1 and 1, in linearly separable data, is represented by $w^T S + b = 0$, where w stands for a weight vector, b indicates the bias, and S is the training sample.

The purpose of the SVM classifier is to reach the optimal values of both w and b to orientate the hyperplane to be as far away as possible from the closest samples and to create the two planes, $H_1 \rightarrow w^T S + b = +1$ for $y_i = +1$ and $H_1 \rightarrow w^T S + b = -1$ for $y_i = -1$, where $w^T S + b \geq +1$ for positive class, and $w^T S + b \leq -1$ for negative class. The combination of these two planes can be expressed as follows: $y_i (w^T S_i + b) - 1 \geq 0 \forall i = 1, 2, 3, \ldots \ldots, M$. The margin of the SVM is represented by the distances d_1 and d_2, which stand for the distances from points H_1 and H_2 to the hyperplane, respectively. Since the hyperplane is located at the same distance from both planes, we can deduce that $d_1 = d_2 = \dfrac{1}{\|w\|}$, and the margin is equal to the total of d_1 and d_2. To get the most out of the margin width, you should do the following:

$$min \frac{1}{2} \|w\|^2 \tag{8.1}$$

$$\text{Subject to } y_i \left(w^T S_i + b \right) - 1 \geq 0 \, \forall i = 1, 2, 3, \ldots\ldots, M \tag{8.2}$$

The quadratic programming problem is represented in Equation 8.2, and the objective function is denoted by the symbol Equation 8.1. The constraint is denoted by Equation 8.2. The following is a formalization of this equation that may be done using the Lagrange formula:

$$\min L_P = \frac{\|w\|^2}{2} - \sum_{i=0}^{M} \alpha_i y_i \left(w^T S_i + b \right) + \sum_{i=0}^{M} \alpha_i \tag{8.3}$$

where α_i are the Lagrange multipliers, each α_i corresponds to one training sample, and L_P refers to the primal problem. Calculating the values of w, b, and α which may be done by differentiating L_P regarding w and b and then assigning the derivatives to zero yields the following results:

$$\frac{\partial L_P}{\partial w} = \sum_{i=1}^{M} \alpha_i y_i S_i = 0 \tag{8.4}$$

$$\frac{\partial L_P}{\partial b} = \sum_{i=1}^{M} \alpha_i y_i = 0 \tag{8.5}$$

By plugging Equations 8.4 and 8.5 into Equation 8.3, the dual problem can be expressed in the following manner:

$$\max L_D = \sum_{i=1}^{M} \alpha_i - \frac{1}{2} \sum_{i,j=1}^{M} \alpha_i y_i \alpha_j y_j S_i^T S_j \tag{8.6}$$

Subject to:

$$\alpha_i \geq 0, \sum_{i=1}^{M} \alpha_i y_i = 0 \, \forall i = 1, 2, 3, \ldots\ldots, M$$

where L_D is an abbreviation for the dual form of L_P. In dual support vector machines, the objective function is optimized to be maximized with respect to α_i rather than to be minimized with respect to w and b. Obtaining the values of w, b, and α can be achieved by solving Equations 8.4, 8.5, and 8.6.

The scenario of non-separable data leads to an increased number of incorrectly categorized patterns. Therefore, to relax the restrictions of the linear SVM, a slack variable called ε_i has been introduced. This variable represents the distance that exists between the *ith* training sample and the associated margin hyperplane, and it needs to be as little as possible. Following the addition of ε_i the objective function of the SVM will be as follows:

$$\min \frac{1}{2} \|w\|^2 + C \sum_{i=1}^{M} \varepsilon_i \tag{8.7}$$

$$\text{Subject to } y_i \left(w^T S_i + b \right) - 1 + \varepsilon_i \geq 0 \, \forall i = 1, 2, 3, \ldots, M \tag{8.8}$$

where C determines how much weight the slack variable penalty should be given in comparison to the size of the SVM margin. The Lagrange equation for Eq. (8.8) can be written as follows:

$$\min L_p = \frac{\|w\|^2}{2} + C \sum_{i=1}^{M} \varepsilon_i - \sum_{i=1}^{M} \alpha_i \left[y_i \left(w^T S_i + b \right) - 1 + \varepsilon_i \right] \tag{8.9}$$

where $\alpha_i > 0$

The value of ε_i can be determined by differentiating L_p with regard to ε_i in the manner shown below:

$$\frac{\partial L_p}{\partial \varepsilon_i} = \sum_{i=1}^{M} \alpha_i y_i S_i = 0 \tag{8.10}$$

Suppose the data cannot be separated linearly. In that case, the SVM will use kernel functions to transform the data into a higher-dimensional space by employing a non-linear kernel function (φ), in which case the data will be able to be linearly separated. The following will serve as the objective function of the SVM:

$$min \frac{1}{2} \|w\|^2 + C \sum_{i=1}^{M} \varepsilon_i \tag{8.11}$$

$$\text{Subject to } y_i \left(w^T \varphi(S_i) + b \right) - 1 + \varepsilon_i \geq 0 \quad \forall i = 1, 2, 3, \ldots, M \tag{8.12}$$

There are a variety of kernel functions, including linear kernel represented according to $\left(K \left(S_i, S_j \right) = \left\langle S_i, S_j \right\rangle \right)$, polynomial kernel of degree represented according to the following formula $K \left(S_i, S_j \right) = \left(\left\langle S_i, S_j \right\rangle + c \right)^2$, and Gaussian kernel, also called as radial basis function (RBF) formulated as follows $K \left(S_i, S_j \right) = exp \left(\frac{\|S_i - S_j\|^2}{2\sigma^2} \right)$ [17]. The

Gaussian kernel function is utilized in the SVM classifier in this chapter in order to find the best or near-optimal solutions.

8.3   ADAPTATION OF METAHEURISTICS FOR PARAMETER ESTIMATION OF SVM

This chapter investigates the performance of seven metaheuristic algorithms: WOA, SMA, MPA, DE, TLBO, GWO, and EO for supplying the SVM classifier with the optimum values of C and the kernel parameter so that it can be trained using the training set. Consequently, the search space was of the two-dimensional variety,

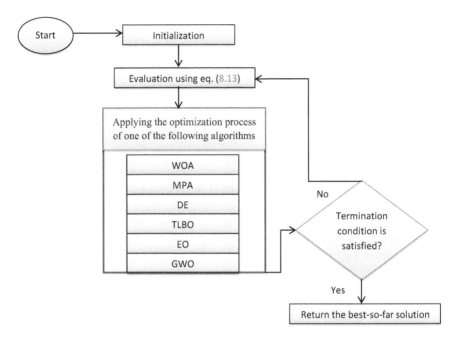

FIGURE 8.4 Flowchart of adapting metaheuristics for parameter estimation of SVM.

and each solution reflects a different combination of the C and kernel parameters. Each investigated metaheuristic algorithm must distribute its solutions within the problem's bounds. The searching range of SVM's parameter C was bounded by C_L = 0.01 and C_{max} = 35,000, and the searching range of σ was bounded by L = 0.01 and H = 100 [18]. After distribution solutions, their fitness values are calculated using the following fitness function:

$$Minimize: F = \frac{N'_e}{N'} \times 100 \qquad (8.13)$$

The SVM classifier is trained using the training set, while the misclassification rate is computed using the testing set. The misclassification rate is defined as the ratio of the number of samples (N'_e) that were incorrectly categorized to the total number of samples (N') that were tested as described in Equation 8.13. The fitness values of all solutions will be compared to extract the best-so-far one X^* for guiding the solutions of each algorithm during the optimization process even reaching the termination condition represented by the maximum number of iterations which is set in this chapter to 100. Figure 8.4 presents the flowchart of adapting metaheuristics for parameter estimation of SVM.

8.4 EXPERIMENT SETTINGS

The effectiveness of seven distinct metaheuristic algorithms that have been described in the past has been assessed by means of a variety of distinct experiments. These

TABLE 8.1
Dataset description

ID	Dataset	No.F	No.S	No.C
ID1	Sonar	60	208	2
ID2	Wisconsin Diagnosis Breast Cancer (WBCD)	32	596	2
ID3	liver-disorders	6	345	2
ID4	Heart diseases	13	270	2

algorithms are WOA, GWO, SMA, EO, DE, TLBO, and MPA, and are evaluated using four well-known datasets taken from the UCI machine learning repository. Table 8.1 contains a listing of all of the datasets' respective descriptions. These datasets are utilized extensively in the process of evaluating the efficacy of various classification models found in published research. K-fold cross-validation tests have been utilized throughout every experiment. In k-fold cross-validation, the dataset's initial samples are randomly divided into k subsets that are (about) the same size, and the experiment is carried out k times. This ensures that the results are reliable. At each time, one of the subsets was selected randomly to serve as the testing set, while the remaining k-1 subsets were utilized as the training set. After that, a single estimation can be produced by simply calculating the average of the k results obtained from the folds. In this chapter, the value of k was determined to be 5. It is impossible to correctly partition the dataset since the number of samples inside each class does not constitute a multiple of 5 (see Table 8.1). On the other hand, the ratio of the number of samples in the training set to the number of samples in the testing set was kept as close to 4:1 as possible.

All of the experiments are carried out using a computer equipped with 32 gigabytes of random access memory (RAM), a 2.40 GHz Intel Core i7–4700MQ CPU, and a Windows 10 operating system. All of the algorithms investigated in this chapter are implemented using MATLAB, with a population size of ten, and a maximum number of iterations of 100. In addition to this, statistical analyses are utilized to validate the obtained findings.

8.5   CHOICE OF PARAMETERS

In this chapter, the results of many tests designed to determine the parameter value that is most suitable for each algorithm in terms of its capacity to solve the parameter estimation of SVM are presented. Extensive experiments under various values for each parameter have been done to achieve this objective for MPA, which has two primary parameters: P and FADs, which need to be accurately estimated to maximize its performance. To achieve this objective, MPA has been subjected to various values for each parameter. The average fitness values gained under these distinct values for each parameter of MPA are depicted in Figure 8.5. When you examine this figure in further detail, you will discover that the ideal value for FADs is 0.2, while the ideal value for P is 5. To get the most out of the EO method, its performance can be increased by accurately forecasting its four essential effective characteristics,

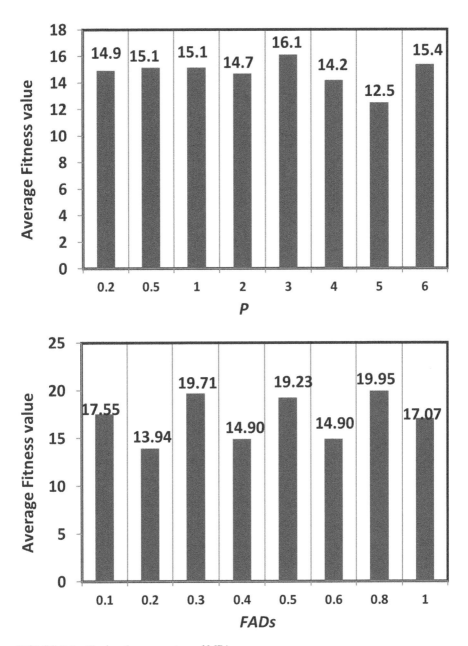

FIGURE 8.5   Tuning the parameters of MPA.

which are indicated by the letters *a1, a2, V, and GP*, respectively. Therefore, several experiments are herein done to investigate a wide range of different possible values for these parameters to find the one that contributes the most to an improvement in the overall performance of EO (see Figure 8.6). As a result of examining this figure, we

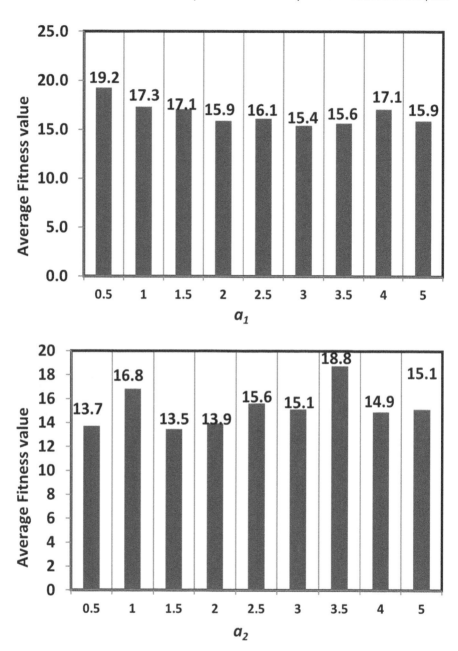

FIGURE 8.6 Tuning the parameters of EO.

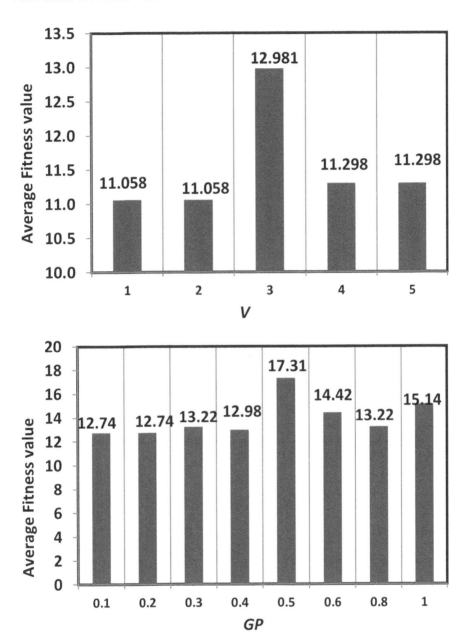

FIGURE 8.6   (Continued)

might get the conclusion that the ideal values for these parameters are, in order, 3, 1.5, 1, and 0.1. Regarding the parameters of DE, which are F and Cr, the ideal values for them are estimated by running DE under a range of possible values for each parameter. The obtained results of these experiments conducted to estimate the parameters of DE are depicted in Figure 8.7.

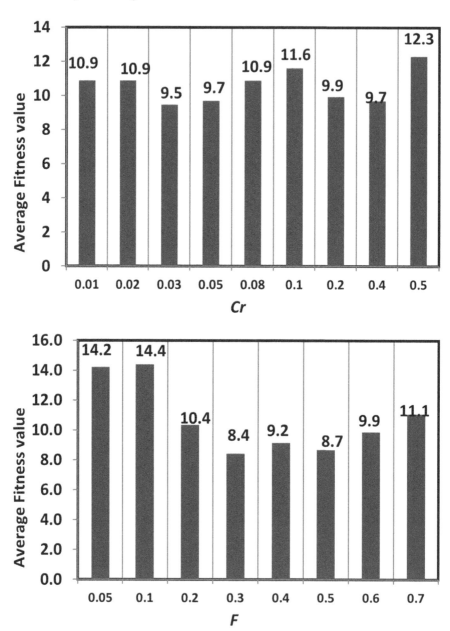

FIGURE 8.7 Tuning the parameters of DE.

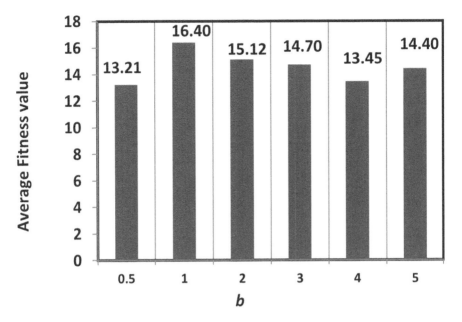

FIGURE 8.8   Tuning the parameter b of WOA.

After analysing the data in this figure, we realized that the optimum values for F and CR, respectively, were 0.3 and 0.03. The WOA only uses one practical parameter, represented by the letter b. This parameter determines the contours of the logarithmic spiral. Experiments have been conducted using a variety of values for this parameter, including 0.5, 1, 2, 3, 4, and 5, amongst others, to determine which of these numbers is the most appropriate. The results of these studies, depicted in Figure 8.8, make it abundantly evident that the number 0.5 is the optimal choice for this parameter under all circumstances. In conclusion, the SMA uses a single effective parameter, denoted by the letter z. This is the only parameter that the SMA utilizes. The results of many trials, each of which utilized a different value for this parameter, are used to determine how effective this parameter is. These results are displayed in Figure 8.9. According to the information that is displayed in this figure, the parameter in question would benefit most from having the value of 0.05 assigned to it.

8.6   RESULTS AND DISCUSSION

Four datasets are employed in this chapter to investigate the performance of seven previously-described metaheuristic algorithms for estimating two parameters of SVM accurately to improve its accuracy for the medical datasets. Every search agent of the metaheuristic algorithms in this study has been executed five independent times, and the average fitness value (F-value) obtained by each algorithm in addition to the standard deviation (SD) on each observed dataset are depicted in Table 8.2. From this table, TLBO could be better than all the other metaheuristics in terms of

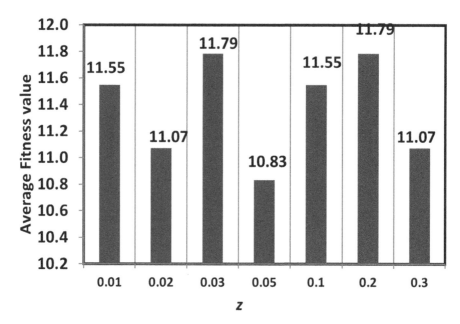

FIGURE 8.9   Tuning the parameter z of SMA.

the average F-value on all datasets and competitive with some of the others for SD. Figure 8.10(a) depicts the average fitness value obtained by each algorithm on four investigated instances, which affirms that TLBO is the best-optimizer for estimating the parameters of SVM to maximizing its accuracy for the medical instances, followed by SMA as the second-best optimizer. Figure 8.10(b) depicts the average SD obtained by each algorithm on all datasets. In addition, Figure 8.11 is presented to depict the convergence speed of each optimizer; this figure demonstrates that TLBO has a faster and better convergence rate than all the other algorithms on all investigated datasets.

8.7   CHAPTER SUMMARY

This chapter addresses the function of metaheuristic algorithms in adjusting two parameters of the support vector machine VM to achieve greater classification accuracy while classifying medical data. Because of this, we chose seven well-known metaheuristic algorithms, each of which was selected based on how recently they were published and how highly they were cited. WOA, DE, MPA, GWO, TLBO, SMA, and EO are the names of these algorithms. After that, these algorithms have been tested through several various tests that have been carried out using four well-known datasets taken from the UCI machine learning repository. The average fitness value, also known as the F-value, and the standard deviation (SD) of each algorithm were employed to compare these algorithms. In conclusion, the results of the experiments demonstrate that the TLBO algorithm has superior performance when compared to all of the other algorithms. Consequently, the TLBO is an effective parameter estimation

TABLE 8.2
Comparison among algorithms in terms of F-value and SD over various datasets

Algorithms	ID1		ID2		ID3		ID4	
	F-value	SD	F-value	SD	F-value	SD	F-value	SD
GWO	10.595238	1.346870	3.950474	0.124054	24.782609	0.204958	18.333333	0.261891
DE	12.059524	3.417683	4.126688	0.125152	24.347826	**0.000000**	18.333333	0.261891
WOA	10.392857	1.060660	4.653780	0.372161	24.202899	1.844626	18.703704	0.785674
MPA	9.642857	**0.000000**	4.215184	**0.000000**	24.347826	**0.000000**	18.148148	**0.000000**
TLBO	**9.404762**	0.336718	**3.336438**	**0.000000**	**23.043478**	0.204958	**16.666667**	**0.000000**
EO	10.119048	0.673435	4.126688	0.125152	24.057971	1.229751	18.703704	0.261891
SMA	9.642857	**0.000000**	4.125912	0.124054	23.188406	0.819834	18.333333	0.261891

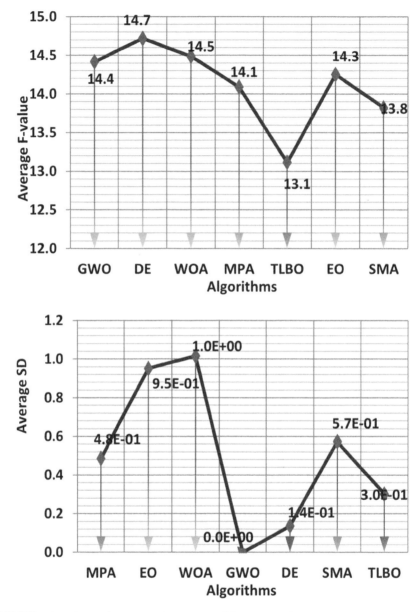

FIGURE 8.10 Comparison under the average of F-value and SD.

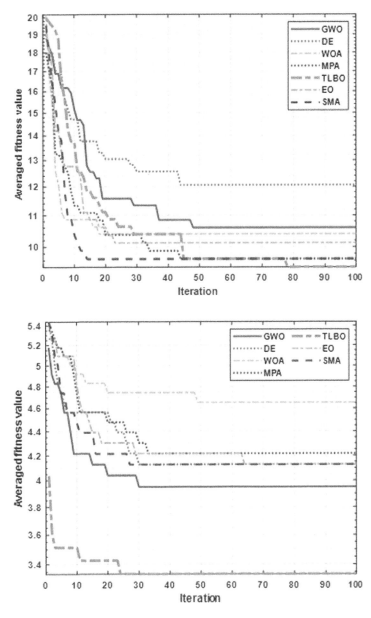

FIGURE 8.11 Comparison among algorithms in terms of the convergence speed on each observed dataset.

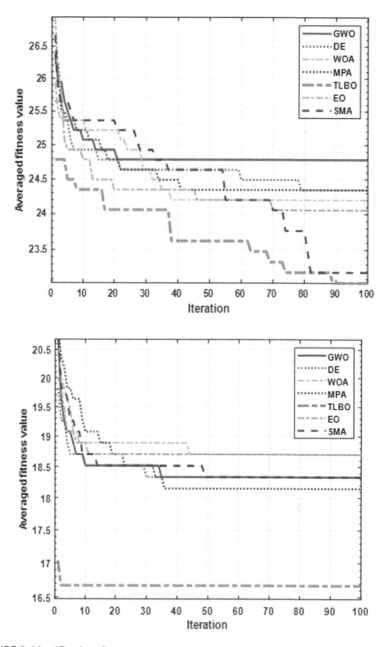

FIGURE 8.11   (Continued)

method for determining the two parameters of the SVM that will lead to improved classification accuracy for medical datasets.

8.8 EXERCISES

a. Define supervised, unsupervised and reinforcement machine learning models. How they are different to each other. Explain with examples.
b. Why is SVM a supervised learning system? Provide an example.
c. What are the parameters than control the performance of SVM? How can we tune them?
d. How can a metaheuristic algorithm influence the performance of SVM for healthcare data?
e. Explain a two-dimensional hyperplane SVM with an example.
f. How can we adapt the metaheuristic algorithm to the parameter estimation of SVM? Provide an example.
g. Why is proper tuning of algorithmic parameters important? How can we do that?

REFERENCES

1. Shafique, U., et al., Data mining in healthcare for heart diseases. *International Journal of Innovation and Applied Studies,* 2015. **10**(4): p. 1312.
2. Almarabeh, H. and E. Amer, A study of data mining techniques accuracy for healthcare. *International Journal of Computer Applications,* 2017. **168**(3): pp. 12–17.
3. Ahmad, A., L.J.D. Dey, A k-mean clustering algorithm for mixed numeric and categorical data. *Data & Knowledge Engineering,* 2007. **63**(2): pp. 503–527.
4. Shmilovici, A., Support vector machines, in *Data Mining and Knowledge Discovery Handbook.* 2009, Springer. pp. 231–247.
5. Zhang, X., X. Chen, and Z. He, An ACO-based algorithm for parameter optimization of support vector machines. *Expert Systems with Applications,* 2010. **37**(9): pp. 6618–6628.
6. Tharwat, A., T. Gabel, Parameters optimization of support vector machines for imbalanced data using social ski driver algorithm. *Neural Computing and Applications,* 2020. **32**(11): pp. 6925–6938.
7. Li, S., H. Fang, and X. Liu, Parameter optimization of support vector regression based on sine cosine algorithm. *Expert Systems with Applications,* 2018. **91**: pp. 63–77.
8. Hassanien, A.E., et al., Intelligent human emotion recognition based on elephant herding optimization tuned support vector regression. *Biomedical Signal Processing and Control,* 2018. **45**: pp. 182–191.
9. Tuba, E., M. Tuba, and M. Beko. Support vector machine parameters optimization by enhanced fireworks algorithm. In *International Conference on Swarm Intelligence.* 2016. Springer.
10. Pereira, D.R., et al., Social-spider optimization-based support vector machines applied for energy theft detection. *Computers & Electrical Engineering,* 2016. **49**: pp. 25–38.
11. Tharwat, A. and A.E. Hassanien, Quantum-behaved particle swarm optimization for parameter optimization of support vector machine. *Journal of Classification,* 2019. **36**(3): pp. 576–598.

12. Li, Q., et al., State of health estimation of lithium-ion battery based on improved ant lion optimization and support vector regression. *Journal of Energy Storage*, 2022. **50**: p. 104215.

13. Xia, J., et al., Performance optimization of support vector machine with oppositional grasshopper optimization for acute appendicitis diagnosis. *Computers in Biology and Medicine*, 2022: p. 105206.

14. Pan, J., Y. Li, and P. Wu. A predict method of water pump operating state based on improved particle swarm optimization of support vector machine. *Journal of Physics: Conference Series*. 2022. IOP Publishing.

15. Hossain, S.Z., et al., Hybrid support vector regression and crow search algorithm for modeling and multiobjective optimization of microalgae-based wastewater treatment. *Journal of Environmental Management*, 2022. **301**: p. 113783.

16. Chapelle, O., et al., Choosing multiple parameters for support vector machines. *Machine Learning*, 2002. **46**(1): pp. 131–159.

17. Tharwat, A. and A.E. Hassanien, Chaotic antlion algorithm for parameter optimization of support vector machine. *Applied Intelligence*, 2018. **48**(3): pp. 670–686.

18. Lin, S.-W., et al., Particle swarm optimization for parameter determination and feature selection of support vector machines. *Expert Systems with Applications*, 2008. **35**(4): pp. 1817–1824.

9 Deep Learning Models to Process Healthcare Data

Introduction

9.1 DEEP LEARNING TECHNIQUES FOR HEALTHCARE

Deep learning (DL) has grown in popularity as a result of the rapid advancement of computer technologies and the exploding expansion of big unstructured data. DL can be classified as a subset of artificial intelligence (AI). AI contains a branch known as machine learning (ML), which employs a set of ML techniques to allow computers to automatically construct a model for complex relationships or patterns based on actual data without having to be explicitly coded. The model for a given task (e.g., classification of localized liver lesions) is trained initially with labelled training inputs and labels. Afterwards, this model can make predictions or perform actions. DL, a subclass of ML, gains greater strength and flexibility than traditional ML models by emulating biological neural networks to perform a wide range of complicated tasks, such as natural language processing (NLP) and medical image classification. In DL, a neural network with several layers between input and output, namely hidden layers, is used. The fundamental advantage of DL is that it can automatically train data-driven, highly representative, and hierarchical features and conduct feature extraction and classification on a single network. DL approaches have achieved considerable success in a variety of computer vision applications, including image classification, detection, and segmentation, and are now playing an essential role in a variety of academic and industrial fields. Recently, DL has been widely applied in medical applications such as anatomic modelling (anatomical structure segmentation), tumour identification, computer-aided diagnosis, disease categorization, and surgical planning. Briefly, the application of DL techniques in the healthcare field has taken place on various scales, ranging from forecasting the trajectory of disease propagation to building diagnostic and prognostic models.

The objective of this chapter is to illustrate the role of metaheuristic algorithms for estimating the hyperparameters of the DL techniques to improve their performance accuracy when tackling healthcare tasks. In general, the organization of this chapter can be summarized as follows. This chapter begins with a discussion on well-known DL techniques like a recurrent neural network, convolutional neural network, and deep neural network to show their hyperparameters which the metaheuristics could estimate to improve their performance. After that, some well-known optimizers employed to update the weights of the DL models are discussed, in addition to

DOI: 10.1201/9781003325574-9

describing some of the activation functions. Next, DL techniques' applications to the healthcare system are discussed. Last but not least, the role of metaheuristics for tuning the hyperparameters of the DL techniques is presented. Finally, a synopsis of the chapter is presented here for your perusal.

9.2   DEEP LEARNING TECHNIQUES

Algorithms based on AI can study past data to forecast the future. Recently, a new branch of ML, DL, has been developed to simulate the human brain more accurately to enable computers to learn, monitor, and respond to complicated situations more quickly than human and classical ML models. There are several ML algorithms: decision tree algorithm (DTA), naïve Bayes, random forest, K-means clustering, support vector machine (SVM), K-nearest neighbour algorithm, and apriori algorithm. The metaheuristic algorithms significantly maximize the performance of most ML algorithms for better accuracy when classifying and clustering datasets of several fields, especially medical fields. For example, metaheuristic algorithms have been widely employed in the K-mean clustering algorithm to estimate the optimal number of clusters in addition to the initial cluster centroids [1]. As mentioned in the previous chapter, the SVM has two effective parameters, C and the kernel parameter, that must be accurately estimated to maximize its performance. Several metaheuristic algorithms have estimated these two parameters as discussed in several studies [2–4].

On the other hand, several DL models based on the deep neural network have been proposed to offer the highest level of accuracy possible in various tasks, including object detection and speech recognition. They can study independently without any predetermined knowledge that the programmers have specifically written. So basically, DL models are just neural networks with numerous hidden layers; some of those models which are so common are the deep neural network, deep belief networks, recurrent neural network, and convolutional neural network. Some of those DL models are discussed in detail in the rest of this section.

9.2.1   Deep Neural Network (DNN)

In its most basic form, a deep neural network (DNN) is an entirely connected feed-forward neural network that has multiple hidden layers. A typical N-layer DNN architecture consists of an input layer, $(N-1)$ hidden layers, and an Nth output layer, as shown in Figure 9.1. Suppose that there is a dataset that has M training samples $\left(x = x_1, x_2, x_3, \ldots\ldots, x_M\right)$, such that x_i refers to the ith training sample with d features. As a result, the training set is represented by the notation $\left(\left(x_1, y_1\right), (x_2, y_2), \left(x_3, y_3\right), \ldots\ldots, (x_M, y_M)\right)$, where $y_1, y_2, \ldots\ldots, y_M$ stand for the class labels associated with x_1, x_2,..., and x_M, respectively. The ith hidden layer is made up of a bias and k units, and there are no inherent connections between any two of the units in any of the layers. On the other hand, every two units of two subsequent layers, i.e. ith and $(i+1)$th, have a weight-based relationship w_{lj}, where w_{lj} indicates the lth and jth units of the ith and jth layers. For the sake of brevity, let's use the notation $I - h_1 - h_2 - h_3 \ldots\ldots - h_{N-1} - O$ to describe this N-layer DNN architecture. For example, the following architecture (6-100-100-100-100-4) indicates that the DNN model

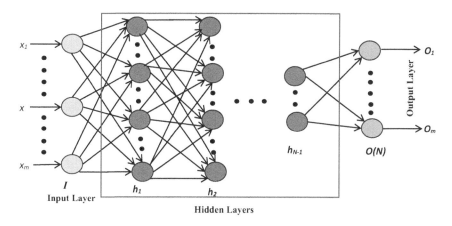

FIGURE 9.1 Deep neural network.

consists of an input layer (I) with 6 neurons, 4 hidden layers (h) with 100 neurons, and an output layer (O) with 4 neurons. The input of each hidden neuron, referred to as the weighted sums of inputs, is computed according to the following formula:

$$s_j = \left(\sum_{l=1}^{l=n} w_{lj}^{i+1} \cdot o_j^i \right) + \theta_j^{i+1} \tag{9.1}$$

where n denotes the number of input neurons, w_{ij} indicates the connection weight between the lth neuron in the ith layer and the jth neuron in the $(i+1)^{th}$ layer, θ_j indicates the bias of the jth hidden neuron in the $(i+1)^{th}$ layer, and o_j^i indicates the outputs of the jth neuron in the ith layer. After that, the output of each hidden neuron o_j^{i+1} in the $(i+1)^{th}$ the layer is computed according to the following formula:

$$o_j^{i+1} = f\left(s_j\right) \tag{9.2}$$

where $f(.)$ indicates an activation function. Several activation functions are proposed for improving the accuracy of the estimated labels by the DL model, which will be discussed later. The objective of the DNN model is to identify a universal approximator that can estimate the optimal label of each input record within the dataset. To do this, all weights are taken into account as design dimensions and optimized so that the most commonly used loss function, also referred to as the objective function, Categorical cross-entropy (CCE), is minimized. Consequently, this loss function can be described mathematically as follows:

$$\text{Min: } loss\left(CCE\right) = -\sum_{i=1}^{M} y_i \cdot \log \breve{y}_i \tag{9.3}$$

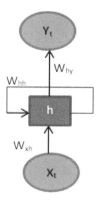

FIGURE 9.2   Fully connected architecture of RNN at time t.

where $\bar{y}_i$ stands for a vector including the predicted outputs of the training samples. This objective function is addressed by several optimization algorithms, like gradient-based algorithms or metaheuristic (stochastic) algorithms.

The gradient-based algorithms have been preferred to minimize the loss functions for reaching the universal approximator because it is faster in reaching the near-optimal weights, which minimizes the loss function and requires a lower computational cost. However, the gradient-based optimizers are prone to falling into local minima. In the next sections, we will discuss some of the well-known gradient-based optimizers like stochastic gradient descent (SGD) algorithm, momentum gradient descent (MGD), adaptive gradient (AdaGrad) algorithm, root mean square propagation (RMSprop), AdaDelta, Adamax, adaptive moment estimation (Adam) algorithm, Nadam, and follow the regularized leader (Ftrl).

9.2.2 Recurrent Neural Network

The recurrent neural network (RNN) is a class of deep neural networks designed to model sequential or time-series data. This RNN contains an input, hidden, and output layer. Additionally, there is a loop from the hidden layer back to the hidden layer; this essential aspect differentiates RNN from other neural networks. This loop allows us to think about sequence difficulties and compels the network to learn the long-term reliance on sequences. RNN is frequently employed for ordinal or temporal issues, such as language translation, natural language processing (NLP), speech recognition, and image captioning; they are integrated into popular apps like Siri, voice search, and Google Translate. Recurrent neural networks, like feed-forward and convolutional neural networks (CNNs), employ training data to learn. The output of all steps at various times (*t-1, t, t+1*) depicted in Figure 9.3 are computed according to the following formula:

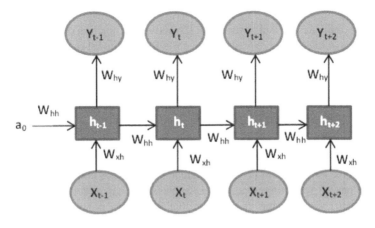

FIGURE 9.3 Fully connected architecture of RNN at different time.

- At the time (t-1):

$$h_{t-1} = W_{xh}X_{t-1} + a_0 W_{hh} + b_h \tag{9.4}$$

$$h_{t-1} = f_h\left(h_{t-1}\right) \tag{9.5}$$

$$Y_{t-1} = f_o\left(W_{hy}h_{t-1} + b_y\right) \tag{9.6}$$

- At time t:

$$h_t = W_{xh}X_t + h_{t-1}W_{hh} + b_h \tag{9.7}$$

$$h_t = f_h\left(h_t\right) \tag{9.8}$$

$$Y_t = f_o\left(W_{hy}h_t + b_y\right) \tag{9.9}$$

- At time t+1:

$$h_{t+1} = W_{xh}X_{t+1} + h_t W_{hh} + b_h \tag{9.10}$$

$$h_{t+1} = f_h\left(h_{t+1}\right) \tag{9.11}$$

$$Y_{t+1} = f_o\left(W_{hy}h_{t+1} + b_y\right) \tag{9.12}$$

where b_h and b_y refer to the bias vectors of the hidden and output layers, respectively. W_{xh}, W_{hh}, and W_{hy} indicates the weighted matrices from input to hidden, from hidden to hidden, and from hidden to output, respectively. The estimated output is represented by Y_t and a_0 indicates. f_h and f_o represent the activation function of hidden and output neurons, respectively. The overall loss J of the RNN is equal to the sum of all of its losses across all timesteps or sequences:

$$J = \sum_{t=1} Y_{t-1}^a - Y_{t-1}^e \qquad (9.13)$$

where Y_{t-1}^a and Y_{t-1}^e indicate the actual and estimated outputs for the sequence t-1, respectively.

RNNs have two gradient problems: vanishing gradients and exploding gradients. The problem of vanishing gradients is a problem that affects RNNs. When the gradients get too small, the information that they carry is no longer useful to the RNN, and this causes the parameter updates to become meaningless. Because of this, learning lengthy data sequences is made more challenging. The phenomenon known as an exploding gradient occurs when the slope of the learning curve tends to rise exponentially rather than diminish. During the training process, this issue manifests itself when huge error gradients develop, which causes unusually large updates to the model weights used by the neural network. The most significant challenges associated with gradient problems include a protracted training period, low performance, and poor precision. The long short-term memory network (LSTMs) discussed in the next section, which is a novel variation of RNN, has proven to be the most effective solution for dealing with gradient difficulties.

9.2.3 LONG SHORT-TERM MEMORY NETWORKS

It is necessary to find a solution to the problem of vanishing and exploding gradients to make it possible for deeper networks and recurrent neural networks to function effectively in real-world contexts. This can be accomplished by finding a way to reduce the multiplication of too small gradients. The LSTM cell is a specially built unit of logic that reduces the vanishing gradient problem sufficiently to make recurrent neural networks more helpful for applications requiring long-term memory, such as text sequence predictions. It accomplishes this by producing an internal memory state that is appended to the processed input, so reducing the multiplicative effect of tiny gradients by a significant amount. An intriguing notion known as a forget gate controls the time dependence and effects of prior inputs by determining which states are remembered or forgotten. LSTM cells also include two more gates, which are referred to as the input gate and the output gate. As stated below, The LSTM consists of four distinct gates with distinct functions: Forget gate (f), input gate (i), input modulation gate (g), and output gate (o), as depicted in Figure 9.4, and the architecture of the LSTM network is shown in Figure 9.5. In this figure, the green circle refers to the element-wise multiplication vector between every two connected gates. The output for the forget gate at time t is defined as follows:

$$f_t = \sigma\left(W_{xf} X_t + h_{t-1} W_{hf} + b_f\right) \qquad (9.14)$$

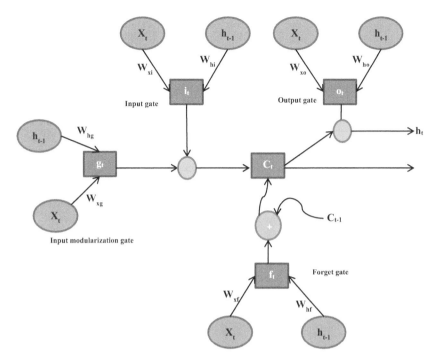

FIGURE 9.4 LSTM recurrent unit.

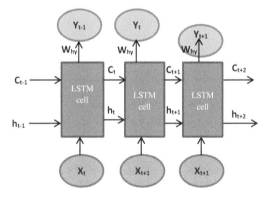

FIGURE 9.5 LSTM network architecture.

Where W_{xf} and W_{hf} are the weights of the forget gate, while b_f stands for the bias of the same gate. σ is the sigmoid activation function. The expression for the input modularization gate is as follows:

$$g_t = tanh\left(W_{xg}X_t + h_{t-1}W_{hg} + b_g\right) \qquad (9.15)$$

where W_{xg} and W_{hg} are the weights of the input modularization gate, while b_g stands for the bias of the same gate. *tanh* is the tanh activation function. The expression for the input gate is as follows:

$$i_t = \sigma\left(W_{xi}X_t + h_{t-1}W_{hi} + b_i\right) \tag{9.16}$$

where W_{xi} and W_{hi} are the weights of the output gate, while b_i stands for the bias of the same gate. The expression for the output gate is as follows:

$$o_t = \sigma\left(W_{xo}X_t + h_{t-1}W_{ho} + b_o\right) \tag{9.17}$$

where W_{xo} and W_{ho} are the weights of the output gate, while b_o stands for the bias of the same gate. The expression for the current internal cell state C_t is as follows:

$$C_t = i_t \otimes g_t + C_{t-1} \otimes f_t \tag{9.18}$$

where $\otimes$ indicates the element-wise multiplication between two vectors. Finally, the output of the LSTM cell can be computed as follows:

$$h_t = o_t \otimes \tanh\left(C_t\right) \tag{9.19}$$

Similar to RNNs, we can apply a linear model over the hidden state h_t and output the likelihood using the softmax activation function to create predictions is computed as follows:

$$Y_t = softmax\left(W_{hy}h_t + b_y\right) \tag{9.20}$$

9.2.4 Convolutional Neural Network

The convolutional neural network (CNN) is a sort of artificial neural network consisting of three layers: convolution, pooling, and fully connected layer. The first two layers, convolution and pooling, are responsible for the feature extraction process. The third layer, known as a fully connected layer, is responsible for mapping the features that have been extracted into the final output, such as classification. These layers essentially feature extractors, dimensionality reduction, and classification layers. These CNN layers are stacked to create a full convolutional layer [5].

Any CNN architecture must have a convolutional layer as its primary focus. This layer consists of a collection of convolutional kernels, often referred to as filters, and is used to build an output feature map by convolving the input image. A kernel is a grid of discrete values or integers, where each value is referred to as the kernel's weight. All weights of a CNN kernel are initialized with random numbers at the beginning of the training phase (different approaches are also available there for initializing the weights). Then, weights are adjusted with each training iteration, and the kernel is taught to extract significant features.

The pooling layers are utilized to sub-sample the feature maps that are created as a result of the convolution operations. This means that it takes the feature maps of a bigger size and reduces them to feature maps of a smaller size. While it is decreasing the size of the feature maps, it will always keep the most important features (or information) in each pool stage. In a manner analogous to the convolution operation, the pooling action is carried out by first providing the size of the pooled region and then the stride of the operation. Different pooling strategies, such as maximum pooling, minimum pooling, average pooling, gated pooling, tree pooling, and so on, are utilized in the various pooling layers. The max pooling approach is by far the most widespread form of pooling. The most significant disadvantage of using a pooling layer is that it might occasionally bring the overall performance of CNN down. The explanation for this is that the pooling layer enables CNN to determine whether or not a certain feature is present in the input image that was provided, regardless of the precise location.

Fully connected layers are often the final layers of every CNN architecture. In these layers, every neuron contained within a layer is coupled with every neuron contained within the layer that came before it. The CNN architecture's output layer (classifier) is the fully connected (FC) layer that comes last in the chain of FC layers. The F) layers receive their input from the last convolutional or pooling layer, which is in the form of feature maps. After that, the maps are flattened to make a vector, which is then fed into the FC layer to provide the final output of the CNN.

9.3 OPTIMIZERS

There are several well-known gradient-based optimizers like stochastic gradient descent (SGD) algorithm, momentum gradient descent (MGD), adaptive gradient (AdaGrad) algorithm, root mean square propagation (RMSprop), AdaDelta, Adamax, adaptive moment estimation (Adam) algorithm, Nadam, and follow the regularized leader (Ftrl), which have been proposed to update the parameters of the Dl and ML models for tackling several real-world problems. In this section, we will discuss the most common optimizer, known as the stochastic gradient descent algorithm, to illustrate how those optimizers could find the local minimum or local maximum of an optimization problem.

Gradient descent is one of the most widely used optimization techniques for training ML and DL models by minimizing the mean of errors between the actual outcomes and the predicted one. The optimization algorithm is the mathematical process of minimizing or maximizing an objective function $f(x)$ parameterized by x. Similarly, in ML, optimization minimizes the errors between the actual and estimated findings as to the cost function by reaching the near-optimal values of the model's weights. There are three different types of gradient descent algorithms: batch gradient descent, mini-batch gradient descent, and stochastic gradient descent

The batch gradient descent algorithm processes all training datasets during each gradient descent iteration. If the number of training datasets is substantial, batch gradient descent will be rather costly. Users are therefore discouraged from employing batch gradient descent if the number of training datasets is significant. For big training datasets, they can instead use mini-batch gradient descent. On the other side, the stochastic gradient descent (SGD) algorithm, in contrast to the batch gradient descent (Batch GD) algorithm, selects a random instance of the training dataset at each iteration to be employed for updating the parameters of the gradient. This results in SGD being significantly faster than Batch GD because there are significantly fewer data to manipulate at any given time. However, SGD will require a huge number of iterations because it processes only one instance at each iteration. Therefore, the mini-batch gradient descent algorithm provides a middle ground between batch gradient descent and SGD by dividing training data into smaller batches. After that, the gradient is computed against a batch for those small batches, and then the parameters of the learning model will be updated based on this gradient. From this, it is observed that the mini-batch gradient descent will update the parameter on a batch of the training dataset rather than only one instance; hence the number of iterations consumed for reaching the local minima is significantly minimized.

The mathematical formula of the gradient descent algorithm to update the parameters is as follows:

$$\theta_j^{+1} = \theta_j - \alpha \nabla_\theta \mathcal{L}\left(\theta;\left(x_i, y_i\right)\right) \tag{9.21}$$

where θ_j^{+1} is a variable to include the updated value of the *jth* dimension, θ_j includes the old value of the same dimension, α stands for the learning rate which could be estimated using the metaheuristic algorithms, $\mathcal{L}\left(\theta;\left(x_i, y_i\right)\right)$ indicates the loss function which is employed to compute the error between the estimated label and the actual one, y_i, while $\nabla_\theta \mathcal{L}\left(\theta;\left(x_i, y_i\right)\right)$ computes the gradient on the training instance $\left(x_i, y_i\right)$.

Suppose that the following cost function needs to be minimized using the SGD to find the optimal values for the variables: θ_1, θ_2 and θ_3:

$$f\left(x_1, x_2\right) = \theta_1 * x_1 + \theta_2 * x_2 + \theta_3 \tag{9.22}$$

Also, the following set of points is given to estimate the near-optimal value for those variables based on them, by minimizing the error value between those given points and the estimated: $f\left(1,1\right) = 5$, $f\left(2,1\right) = 7$. The mean square error is used as a loss function, which is represented as follows for optimizing the previous function:

$$\mathcal{L}\left(\theta;\left(x_1, x_2, y_i\right)\right) = \frac{1}{n}\sum_{i=1}^{2}\left(y_i - \left(\theta_1 * x_1 + \theta_2 * x_2 + \theta_3\right)\right)^2 \tag{9.23}$$

where n indicates the number of samples, which is 2 in this example. Starting with computing the gradient based on computing the derivative of the Equation (9.23) using the partial derivatives because there are three coefficients, and hence each derivative corresponds to its coefficient. For the coefficient θ_1, θ_2, and θ_3, respectively, the partial derivatives are of:

$$\nabla_{\theta_1}\mathcal{L}\left(\theta;\left(x_1,x_2,y_i\right)\right) = -2\frac{1}{n}\sum_{i=1}^{2}\left(y_i - \left(\theta_1 * x_1 + \theta_2 * x_2 + \theta_3\right)\right) * x_1 \qquad (9.24)$$

$$\nabla_{\theta_2}\mathcal{L}\left(\theta;\left(x_1,x_2,y_i\right)\right) = -2\frac{1}{n}\sum_{i=1}^{2}\left(y_i - \left(\theta_1 * x_1 + \theta_2 * x_2 + \theta_3\right)\right) * x_2 \qquad (9.25)$$

$$\nabla_{\theta_3}\mathcal{L}\left(\theta;\left(x_1,x_2,y_i\right)\right) = -2\frac{1}{n}\sum_{i=1}^{2}\left(y_i - \left(\theta_1 * x_1 + \theta_2 * x_2 + \theta_3\right)\right) \qquad (9.26)$$

After that, the SGD will be fired to search for the best value for the parameters: θ_1, θ_2, and θ_3 based on the partial derivatives shown before and the given set of points. For example, at the beginning of the optimization process, three random values of 1, 2, and 3 are assigned to those parameters. Then the predicted outcomes based on those values and the given set of points are computed as follows:

$$f(1,1) = \theta_1 * 1 + \theta_2 * 1 + \theta_3 = 1*1 + 2*1 + 3 = 6 \qquad (9.27)$$

$$f(2,1) = \theta_1 * 2 + \theta_2 * 1 + \theta_3 = 1*2 + 2*1 + 3 = 7 \qquad (9.28)$$

The estimated value for $x_1 = 1$ and $x_2 = 1$ is 6, while the actual value is 5. When $x_1 = 2$ and $x_2 = 1$, the estimated value is 7, while the actual value is 7. Since there is an error rate between the actual value and the estimated one when $x_1 = 1$ and $x_2 = 1$, the coefficient θ_1, θ_2, and θ_3 have to be updated using Equation (9.21) to minimize this error.

9.4 ACTIVATION FUNCTIONS

In a neural network, the activation function specifies how the input is translated into an output from a node or nodes in a network layer. The input's weighted sum is based on the activation function. In some contexts, the activation function is sometimes referred to as a transfer function. Many activation functions exhibit nonlinear behaviour, which may be referred to as nonlinearity when discussing the layer or network architecture. The choice of activation function has a significant bearing on the capabilities and performance of the neural network. Several activation functions may be utilized in various components of the model. This is because different activation functions are best suited for specific tasks. Although, technically speaking, the activation function is employed within or after the internal processing of each node in the network. Networks are designed to use the same activation function for all nodes contained within a single layer.

A network could have three different layers: input layers, which receive unprocessed data from the domain; hidden layers, which receive data from one layer and pass it on to another layer; and output layers, which provide a prediction. Usually, a single activation function is utilized across all hidden layers. The activation function used by the output layer will typically be different from that used by the hidden layers. This difference is determined by the type of prediction that the model requires. Activation functions are often differentiable, which means that the first-order derivative can be computed for a given input value. This property allows activation functions to be used in mathematical analysis. This is necessary because neural networks are often trained by employing the backpropagation of error procedure, which calls for the derivative of prediction error to update the model's weights. Although neural networks can use a wide variety of activation functions, in reality, only a tiny subset of those functions are employed for the hidden and output layers. Some of those activations are discussed in the next sections.

9.4.1 RECTIFIED LINEAR UNITS (RELU)

The rectified linear activation function offers a solution to disappearing gradients, enabling models to learn more quickly and carry out their tasks more effectively. The rectified linear activation is the function used by default when creating DNN and CNN. The mathematical model of this function is as follows:

$$ReLU(x) = max(0, x) \tag{9.29}$$

Therefore, the output of ReLU is 0 when the input has a negative value, and it is x when the input has a positive value. Figure 9.6 depicts the ReLU activation function and its derivative (der_ReLU).

9.4.2 SCALED EXPONENTIAL LINEAR UNIT (SELU)

The central limit theorem forms the basis for the normalization that is incorporated into SELU. The SELU function is an example of a monotonically growing function; for huge amounts of negative input, it produces a negative value that remains essentially constant. SELUs are most frequently implemented in self-normalizing networks (SNNs). The mathematical model of this function is as follows:

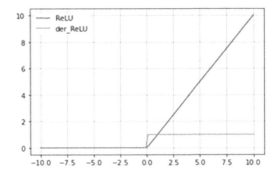

FIGURE 9.6 Depiction of ReLU activation function and derivative.

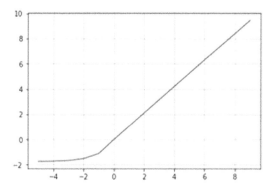

FIGURE 9.7   Depiction of SeLU activation function.

$$SELU(x) = \begin{cases} \lambda x & if \ x > 0 \\ \lambda\alpha(e^x - 1) & if \ x \leq 0 \end{cases} \qquad (9.30)$$

Where λ and α have the following approximate values, respectively: 1.0507 and 1.6732. Figure 9.7 depicts the SeLU activation function.

9.4.3   EXPONENTIAL LINEAR UNIT (ELU)

This activation function addresses some of the issues that have been occurring with ReLUs while preserving some of its desirable aspects. If the x-value you input is greater than zero, then it's the same as the ReLU; in this case, the output will be a y-value equal to the x-value you input. However, if the inputted value x is less than zero, we will obtain a somewhat negative value. The mathematical model of this function is as follows:

$$ELU(x) = \begin{cases} x & if \ x > 0 \\ \alpha(e^x - 1) & if \ x \leq 0 \end{cases} \qquad (9.31)$$

where α is a scalar value and recommended to be picked between 0.1 and 0.3. Figure 9.8 depicts the ELU activation function and its derivative under various values for the parameter.

9.4.4   HYPERBOLIC TANGENT (TANH)

This activation function is extremely similar to the sigmoid function and even has the same S-shaped curve as shown in Figure 9.9. Any real value can be used as an input for the function, and it will return values in the range from −1 to 1. If the value of the input is greater, then the value of the output will be closer to 1.0. On the other hand, if

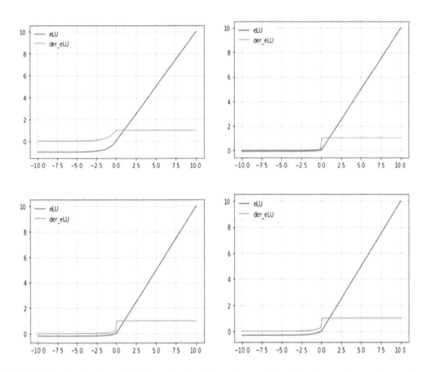

FIGURE 9.8 Depiction of ELU activation function and its derivative under various values for α.

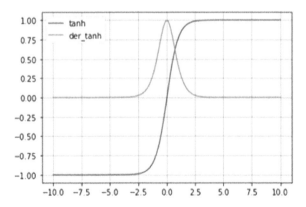

FIGURE 9.9 Depiction of hyperbolic tangent activation function and its derivative under various values for α.

the value of the input is smaller, then the output will be closer to -1.0. The mathematical model of this function is as follows:

$$tanh(x) = \frac{\sinh(x)}{\cosh(x)} = \frac{e^x - e^{-x}}{e^x + e^{-x}} \tag{9.32}$$

The derivative of this activation function is computed as follows:

$$tanh'(x) = \frac{d}{dx}\frac{\sinh(x)}{\cosh(x)} = \frac{\frac{d}{dx}\sinh(x)\cosh(x) - \frac{d}{dx}\cosh(x)\sinh(x)}{\cosh^2(x)}$$

$$= \frac{\cosh^2(x) - \sinh^2(x)}{\cosh^2(x)} = 1 - \frac{\sinh^2(x)}{\cosh^2(x)} = 1 - \tanh^2(x) \tag{9.33}$$

9.5 ROLE OF DEEP LEARNING MODELS IN THE HEALTHCARE SYSTEM

DL has inaugurated a new era of study. It has been utilized in a variety of sectors, one of which, health informatics, has made significant strides over the past few years. Applying DL models in the medical field will result in beneficial outcomes. Because these models take into account a variety of characteristics of the patients' data, such as the variations in molecular traits, diagnostic medical images, elements associated with the environment, electronic health records, and lifestyle. As mentioned in the previous section, there are several DL models, so we will review some of the applications of those models for healthcare.

Both CNN and the deep neural network have been designed to predict and diagnose Parkinson's disease from the medical images [6]. In addition, the CNN was adapted to deal with unbalanced chest X-ray images for improving the accuracy when classifying multiple tuberculosis epidemics by a large margin [7]. The LSTM neural network has been widely applied to several problems in the healthcare system to detect different diseases. For example, the LSTM trained using two nature-inspired metaheuristic algorithms has been developed to predict patients infected with diabetes or cancer [8]. In addition, An accurate method for missing data prediction has been developed for use in the healthcare system that is based on LSTM deep learning [9]. There are several other applications for LSTM in the healthcare system [10–12].

9.6 METAHEURISTIC ROLES FOR HYPERPARAMETERS

After providing a quick overview of the principles of RNN, CNN, LSTM, and DNN, it is not difficult to comprehend the idea of hyperparameters and their significance. During the learning process by any DL technique, the network parameters, such as weights or biases, are assessed and modified. On the other hand, a hyperparameter is a variable whose value is determined before conducting the training process; hence, it does not undergo evaluation or correction. In other words, hyperparameters are the factors that define how the network is trained, like learning rate and momentum rate, as well as the variables that determine the topology of the network, such as the number of hidden layers, optimizer, and activation function, and a number of neurons for each layer. Therefore, hyperparameters are of utmost significance because the performance of the DL technique is heavily dependent on them. The search for accurate values for those hyper-parameters is a difficult operation that is typically carried out by hand and necessitates a significant amount of effort. Due to the significant success achieved by the metaheuristics in several fields, the researchers move toward applying them to estimate the hyperparameters of the DL techniques to maximize their performance for several real-world tasks. In [13–18], several metaheuristic algorithms have been applied to estimate the hyperparameters of CNN, DNN, RNN, and LSTM. In the next chapter, seven previously-described metaheuristic algorithms will be investigated to show which one could optimize the hyperparameters of a deep neural network in order to detect COVID-19 disease from chest X-Ray images.

9.7   CHAPTER SUMMARY

This chapter explains how metaheuristic algorithms can be used to estimate the hyperparameters of DL approaches to improve their performance accuracy when solving healthcare tasks. Hyperparameters are variables that dictate how the network is trained, such as learning rate and momentum rate, as well as variables that establish the network's structure, such as the number of hidden layers, optimizer, and activation function, and a number of neurons for each layer. As a result, hyperparameters are extremely important because the DL technique's performance is largely dependent on them. Therefore, in this chapter, some of the DL techniques like a convolutional neural network, recurrent neural network, and deep neural network are described to illustrate the form of hidden layers for each one. In addition, the hidden layers contain an activation function that specifies how the input is translated into an output from a node or nodes. Therefore, some of the well-known activation functions investigated in the next chapter to define the most suitable one for the deep neural network are discussed here to illustrate the mathematical model and shape of each one. The training process of the DL process is based on an optimizer that updates the parameters, known as weights, in the hope of reaching better accuracy, even reaching the termination condition. Also, this chapter explains some of the gradient-based optimizers as the well-known type of optimizers employed for training the neural network. The next chapter will investigate those optimizers to estimate the most effective one for the DNN models.

9.8   EXERCISES

   a. How do you distinguish between machine learning and deep learning (DL)? Why do we need DL models to process healthcare data?
   b. How can metaheuristic algorithms influence the estimate of the hyperparameters of the DL techniques? Provide an example.
   c. Provide a few examples of deep neural networks. Why do we need a deep neural network to deal with healthcare data?

d. What is a recurrent neural network? How is that different from a deep neural network?

e. Explain the basic architecture of an LSTM model. When should we use LSTM models?

f. How can metaheuristic algorithms augment DL models? Provide an example.

g. Explain the role of DL models in the healthcare system.

REFERENCES

1. Xiao, J., et al., A quantum-inspired genetic algorithm for k-means clustering. *Expert Systems with Applications,* 2010. **37**(7): pp. 4966–4973.

2. Tuba, E., M. Tuba, and D. Simian. Adjusted bat algorithm for tuning of support vector machine parameters. In *2016 IEEE Congress on Evolutionary Computation (CEC).* 2016. IEEE.

3. Korovkinas, K., et al., Support vector machine parameter tuning based on particle swarm optimisation metaheuristic. *Nonlinear Analysis: Modelling and Control,* 2020, **25**(2): pp. 266–281.

4. Stoean, R., et al., *A Support Vector Machine-Inspired Evolutionary Approach for Parameter Tuning in Metaheuristics.* 2009, Citeseer.

5. Bengio, Y., *Learning Deep Architectures for AI.* 2009, Now Publishers Inc.

6. Kollias, D., et al., Deep neural architectures for prediction in healthcare. *Complex & Intelligent Systems,* 2018. **4**(2): pp. 119–131.

7. Liu, C., et al. TX-CNN: Detecting tuberculosis in chest X-ray images using convolutional neural network. In *2017 IEEE international conference on image processing (ICIP).* 2017. IEEE.

8. Rashid, T.A., et al., Improvement of variant adaptable LSTM trained with metaheuristic algorithms for healthcare analysis, in *Advanced Classification Techniques for Healthcare Analysis.* 2019, IGI Global. pp. 111–131.

9. Bhavya, S. and A.S. Pillai. Prediction models in healthcare using deep learning. In *International Conference on Soft Computing and Pattern Recognition.* 2019. Springer.

10. Gao, J., et al., An effective LSTM recurrent network to detect arrhythmia on imbalanced ECG dataset. *Journal of Healthcare Engineering,* 2019: pp. 1–10.

11. Islam, M.S., et al. Intelligent healthcare platform: Cardiovascular disease risk factors prediction using attention module based LSTM. In *2019 2nd International Conference on Artificial Intelligence and Big Data (ICAIBD).* 2019. IEEE.

12. Chimmula, V.K.R., and L.J.C. Zhang, Time series forecasting of COVID-19 transmission in Canada using LSTM networks. *Chaos, Solitons & Fractals,* 2020. **135**: p. 109864.

13. Gaspar, A., et al., Hyperparameter optimization in a convolutional neural network using metaheuristic algorithms, in *Metaheuristics in Machine Learning: Theory and Applications.* 2021, Springer. pp. 37–59.

14. Nematzadeh, S., et al., Tuning hyperparameters of machine learning algorithms and deep neural networks using metaheuristics: A bioinformatics study on biomedical and biological cases. *Computational Biology and Chemistry,* 2022. **97**: p. 107619.

15. Srivastava, D., Y. Singh, and A. Sahoo. Auto tuning of RNN hyper-parameters using cuckoo search algorithm. In *2019 Twelfth International Conference on Contemporary Computing (IC3).* 2019. IEEE.

16. Ahyar, L.F., S. Suyanto, and A. Arifianto. Firefly algorithm-based hyperparameters setting of DRNN for Weather Prediction. In *2020 International Conference on Data Science and Its Applications (ICoDSA).* 2020. IEEE.

17. Aufa, B.Z., S. Suyanto, and A. Arifianto. Hyperparameter setting of LSTM-based language model using grey wolf optimiser. In *2020 International Conference on Data Science and Its Applications (ICoDSA)*. 2020. IEEE.

18. Bouktif, S., et al., Multi-sequence LSTM-RNN deep learning and metaheuristics for electric load forecasting. *Energies*, 2020. **13**(2): p. 391.

10 Metaheuristics to Augment DL Models Applied for Healthcare System

10.1 DL FOR DETECTING COVID-19: HEALTHCARE SYSTEM

The pandemic of the coronavirus disease (COVID-19) and the related efforts to contain it have created a health crisis on a global scale that affects every aspect of human existence. Some of the symptoms of COVID-19 include coughing, fever, and an illness that affects the respiratory system in some people. COVID-19 can even develop into pneumonia [1]. On the other hand, these symptoms do not always imply COVID-19 and are seen in many pneumonia cases, leading to diagnostic challenges for doctors [2]. In general, pneumonia is an infection that inflames the air sacs in the lungs responsible for oxygen transport. Various other causes of pneumonia are fungus, bacteria, and other viruses. Chronic disorders such as bronchitis or asthma, an impaired or weakened immune system, smoking, and ageing contribute to the severity of COVID-19 [1, 3, 4].

In the COVID-19 pandemic, the early identification and diagnosis of COVID-19 and the correct separation of non-COVID-19 cases at the lowest cost and in the early stages of the disease are among the greatest hurdles [5]. Radiological images, such as X-rays and computed tomography (CT) scans, can be extremely helpful in diagnosing COVID-19 as one of the essential methods. Imaging of the chest is a simple and rapid treatment that is suggested by medical and health standards and has been listed in various texts as the primary screening tool during epidemics [2]. In addition, there is another test for diagnosing COVID-19, known as RT-PCR; however, it is an extremely time-consuming, expensive, complex, and manual process. CT scans show a higher sensitivity than RT-PCR for identifying and detecting instances of COVID-19 but lower specificity. This indicates that CT scans are more accurate for COVID-19 but less reliable for non-viral pneumonia [2].

The CT scan's low specificity can lead to difficulties in detecting cases not associated with COVID-19. In addition, the CT scanner's radiation can pose complications for individuals who require many CT scans during a disease. Imaging with X-rays, as opposed to CT scans, requires less rare and costly equipment, which means that significant cost reductions can be made in the operational expenses. In addition, it is possible to employ portable CXR devices in isolated rooms to reduce the danger of infection caused by the usage of these devices in healthcare facilities

[2]. The radiologist is unable to detect pleural effusion with X-rays and cannot calculate the volume involved either.

Numerous studies on the application of deep learning (DL) to the examination of radiological images have been carried out to circumvent the shortcomings of the COVID-19 diagnostic tests that rely on radiological images. Over the last two years, the researchers have moved toward the metaheuristic algorithms to improve the accuracy of the deep learning model for diagnosing COVID-19 disease. For example, in [6], a deep COVID-19 screening model based on the strength of Pareto evolutionary algorithm-II (SPEA-II) has been proposed for X-ray images. This model is based on the modified AlexNet architecture which is used for feature extraction and classification of the input images with the hyper parameters estimated by SPEA-II. There are several other deep learning models based on metaheuristic algorithms to improve its accuracy, like metaheuristics-based COVID-19 detection [7], optimizing the weights of convolutional neural networks by metaheuristics [8], tuning hyper parameters of deep neural networks using metaheuristics [9–12], and several else [13–16].

This chapter's objective is to analyse the efficacy of seven alternative metaheuristic methods for optimizing the hyperparameters and structural design of DNN in the interest of detecting COVID-19 disease from chest X-ray images. To be more specific, a group of seven metaheuristic algorithms called GWO, DE, WOA, MPA, SMA, TLBO, and EO is utilized to estimate the number of layers, the number of neurons that make up each layer, the most effective learning, and momentum rate, as well as the most effective optimizer and activation function of DNN to improve its performance when it comes to detecting COVID-19 disease. This was done to enhance the classification performance of the metaheuristic-based DNN models, namely GWO DNN, DE DNN, WOA DNN, MPA DNN, SMA DNN, TLBO DNN, and EO DNN, for detecting COVID-19 disease from chest X-ray images.

In a general sense, the following is how this chapter is structured: At the beginning of this chapter, the adaptation of metaheuristic algorithms to tune the hyperparameters, estimate the best-performed optimizer, and the best activation function of the deep neural network is discussed. Then, the dataset used to train the models will be described in detail, followed by the section which presents the performance metrics employed to evaluate the solution's performance obtained by each metaheuristic. Last but not least, experimental results based on various performance indicators are presented to demonstrate the efficacy of six researched metaheuristics. Finally, a summary of the chapter is provided.

10.2 ADAPTATION OF METAHEURISTICS FOR TUNING HYPERPARAMETERS OF DNN

This chapter investigates the performance of seven metaheuristic algorithms: WOA, SMA, MPA, DE, TLBO, GWO, and EO, for tuning the hyper-parameters of the deep neural networks (DNNs) to maximize their performance when predicting the COVID-19 infection. Therefore, this section is presented to illustrate how to adapt the metaheuristics for that. The main steps of this adaptation are initialization, evaluation, and finally, the pseudocode of adapted algorithms.

10.2.1 INITIALIZATION PROCESS

When implementing the metaheuristic algorithms, the first stage is creating a number, N, of solutions, referred to as population, where each solution consists of the dimensions of an optimization problem. For example, if we need to find the smallest value of a mathematical function having six variables, then a solution with six dimensions will be created. The optimization problem tackled in this chapter has a number of dimensions which is dynamically changed according to the number of layers in DNN. This number is estimated in this chapter as follows: the number of layers is represented in a dimension (Z); the number of dimensions (M) for the neurons is equal to the number of layers, where all dimensions contain the number of neurons for all layers, respectively; two dimensions (L, Q) for the learning rate and momentum rate; in addition to adding two dimensions, namely O and P, for the optimizer and activation function. Finally, the number of dimensions (d) for each solution to estimate the most effective parameters of DNN is calculated according to the following formula:

$$d = Z + M + O + P + L + Q \qquad (10.1)$$

Since each of the following symbols: $Z, O, P, Q,$ and L need only one dimension, Equation (10.1) is replaced by the following one:

$$d = M + 5 \qquad (10.2)$$

where $M = Z$ represents the number of layers estimated by the updated solution. The solution representation for this problem is shown in Figure 10.1. Here, This dimension represents the number of neurons for each layer; for example, the value of index 5 represents the number of neurons for layer 0, the index 6 represents the number of the neurons for the layer 1, and so on.

After illustrating how to compute the number of dimensions for each solution according to the estimated number of layers, we need to show how these dimensions will be initialized within the search space of each dimension. First, each dimension's upper and lower values have to be identified. Both the learning and the momentum rates have an upper value of 1 and a lower value of 0. However, the upper value and lower value of the learning rate could not be 1 and 0, respectively. In addition, the more the learning rate is near 1, the more the near-optimal solution is skipped because the updated steps are so large.

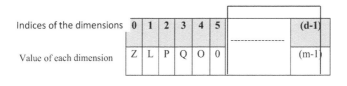

FIGURE 10.1 The solution representation for tuning DNN.

On the contrary, when the value of the learning rate is near 0, the desired solution will need a huge number of epochs and might be intractable. Therefore, our experiments set the lower and upper values for the learning rate to 0.00001 and 0.2, respectively, to avoid the shortcomings. Regarding the number of layers, we set its upper and lower values to 20 and 2, respectively. The researchers could increase these values to increase the chance of reaching a better number of layers, but that will increase the search process time because the algorithms will need to explore a huge number of regions. In relative to the upper and lower values of the number of neurons, we set them to 80 and 2000, respectively. Also, the upper value for the number of neurons could be increased in the hope of finding better solutions, but that will increase the exploration process of the metaheuristic algorithm. Therefore, we preferred to use the previously mentioned upper value (2000) to reduce the computational cost.

Last but not least, nine optimizers described in the previous chapter are investigated by the metaheuristic algorithms in the hope of finding one that could reach a better loss value with the estimated learning rate. Finally, the metaheuristic algorithm investigates four activation functions to find the most effective one, where the upper and lower are 4 and 1, respectively. If the metaheuristic algorithm estimates a value of 1 for the activation function, then, according to Table 10.1, the ReLU activation function will be employed with the other estimated parameters. In addition, the upper and lower values for the optimizers are integer numbers between 1 and 9, where each integer within this interval will represent an optimizer extracted from Table 10.1. Generally, the upper and lower values for the dimensions of this optimization problem are listed in Table 10.3.

The optimizer, activation function, number of layers, and number of neurons have to be integers, and the updated solutions produced by the investigated metaheuristic algorithms might involve decimal values for these parameters due to the nature of the investigated metaheuristic algorithms. So, these updated solutions are first repaired by converting the dimensions of those parameters into an integer to become feasible to tackle this problem. For example, suppose that an optimization algorithm generates

TABLE 10.1
Investigated Optimizers by the Metaheuristic Algorithms

Code	1	2	3	4	5	6	7	8	9
Optimizers	SGD	Adam	RMSprop	Adadelta	Adagrad	Adamax	Nadam	Ftrl	Momentum

TABLE 10.2
Investigated Activation Function

Code	1	2	3	4
Activation functions	relu	selu	elu	tanh

TABLE 10.3
Upper and Lower Values for Each Dimension in This Optimization Problem

Dimensions	Z	L	P	Q	O	M
Upper $\vec{ub}$	2	0.00001	1	0	1	80
Lower $\vec{lb}$	20	0.2	4	1	9	2000

TABLE 10.4
An Unsuitable Solution Produced by an Algorithm for Tuning DNN

0	1	2	3	4	5		(d-1)
5.5	0.2	2.5	0.1	8.6	100.5	-------------------	200.5

TABLE 10.5
The Repaired Solution for Tuning DNN

$\bar{0}$	1	2	3	4	5		(d-1)
5	0.2	2	0.1	8	100	-------------------	200

the solution depicted in Table 10.4 for this problem, but this solution involves infeasible decimal values for the optimizer, activation function, number of layers, and number of neurons which require integer values. To make this solution feasible for tuning the DNN, those decimal values have to be converted into an integer by truncating the decimal numbers after the dot, as shown in Table 10.5.

Now, let's describe how the metaheuristic algorithms will initialize their solutions within the upper and lower bound of each dimension, in the beginning, each algorithm will create N solutions with d dimensions for each one. Then, those solutions will be randomly distributed within the feasible area of each dimension according to the following formula:

$$\vec{X} = \vec{lb} + \vec{r_1} \cdot \left(\vec{ub} - \vec{lb} \right) \qquad (10.3)$$

where $\vec{r_1}$ is a vector including numerical values generated randomly between 0 and 1. As those solutions involve decimal values for some dimensions which must be an integer to be suitable for solving this problem, they are repaired by converting the values inside these dimensions into an integer as described before. After distributing and repairing the initialized solutions, they will be evaluated to get the one, namely the best-so-far solution $\vec{X^*}$, which could reach better values which helps the DNNs find better loss values. Generally, the evaluation process is described in detail in the next section. Algorithm 10.1 describes the initialization process of the metaheuristic algorithms for tackling this problem.

Algorithm 10.1 Initialization

  Return: $\vec{X}$
1. **for** each $\vec{X}_i$ solution
2. $\vec{X}_i$: Create a numerical vector according to Eq. (10.3)
3. **end for**

10.2.2 Evaluation

In the previous chapter, we clarified that the DNN would employ the loss function called categorical cross-entropy (CCE) to be optimized by the estimated optimizer by each metaheuristic algorithm for reaching the weights which fulfil better classification performance for the DNN. Therefore, the solutions generated by the metaheuristic algorithms will be used to construct the DNN, and then this DNN will be trained by the selected optimizer and this loss function. After completing the training process, the loss values obtained under these solutions will be returned to the metaheuristic algorithm to select the solution which could reach the best loss value, the lowest value. This solution will be stored to guide the other solutions within the rest of the optimization process to find better solutions.

10.2.3 Adapting Metaheuristics for DNN

In this section, we will illustrate the pseudocode of one randomly picked metaheuristic, whale optimization algorithm (WOA), after adaptation to tackle the problem of tuning the parameters of the DNN. The other metaheuristic algorithms will be adapted using the same pattern. The WOA has been developed for tackling the continuous optimization problem, and could successfully tackle several optimization problems. Therefore, in this chapter, we will investigate its performance when adapting the DNN for detecting the chest X-ray image infected by COVID-19. The optimization process of the WOA tries to update the initialized solutions to reach a better one. Therefore, before starting the optimization process, a number of N solutions will be generated randomly according to Algorithm 10.1; this initialization step is described earlier as the first step in the pseudocode of WOA adapted for optimizing the DNN (WOA_DNN). After that, the current iteration will be set to 1, and the optimization process will repeatedly continue even satisfying the termination condition, which is based on satisfying the maximum iteration. Finally, after satisfying the termination condition, the best-so-far solution $\vec{X}^*$ will be returned to construct the final DNN model, namely WOA_DNN for detecting the chest X-ray image (the steps are discussed in Algorithm 10.2). Figure 10.2 presents the flowchart of adapting any metaheuristic for tuning the parameters of DNN. The other DNN models structured by GWO, MPA, DE, SMA, EO, and TLBO are called GWO_DNN, MPA_DNN, DE_MPA, SMA_DNN, EO_DNN, and TLBO_DNN, respectively.

Algorithm 10.2 The pseudocode of WOA_DNN

Output: $\overrightarrow{X^*}$

4.   Calling Algorithm 10.1
5.   Find the best whale $\overrightarrow{X^*}$
6. $t = 1$
7.   **while** ($t < T$)
8.   **for** each i whale
9.   Update a, A, p, C, and l
10.   **if** ($p < 0.5$)
11.   **if** ($|A| < 1$)
12.   Update $\overrightarrow{X_i}(t+1)$ using Eq. (1.1)
13.   **else**
14.   Update $\overrightarrow{X_i}(t+1)$ using Eq. (1.8)
15.   **end if**
16.   **else**
17.   Update $\overrightarrow{X_i}(t+1)$ using Eq. (1.6)
18.   **end if**
19.   Check the search boundary of $\overrightarrow{X_i}(t+1)$
20.   Repair $\overrightarrow{X_i}(t+1)$ to avoid decimal values for some dimension
21.   Construct the DNN according to the repaired solution $\overrightarrow{X_i}$
22.   Training the DNN for 5 epochs
23.   Return the loss value obtained by the trained DNN
24.   Replacing the best whale $\overrightarrow{X^*}$ with $\overrightarrow{X_i}(t+1)$
25.   if the loss value obtained by this solution is better.
26.   **end for**
27. t ++
28.   **end while**

10.3 DATASET DESCRIPTION

The dataset used in this study is a composite of two separate datasets that are both available to the public. This dataset includes chest X-ray images taken from patients who had confirmed cases of COVID-19 disease, common bacterial pneumonia, and normal occurrences (no infections). More specifically, the COVID-19, normal, and pneumonia (bacterial) chest X-Ray datasets were collected from [17] and Kaggle's repository. The COVID-19 dataset includes 112 chest X-ray images of lung tissue, and the normal dataset contains 1575 chest X-ray images, while the pneumonia dataset consists of 4265. In a nutshell, the dataset that was utilized for this study has an even distribution of cases in terms of the total number. It comprises three classes (covid, pneumonia, and normal). Some restrictions should be brought to your attention. To begin, the number of confirmed COVID-19 samples already exists is quite low compared to the number of normal or pneumonia cases. In addition, as said in [17], the pneumonia samples are older recorded samples and do not represent pneumonia images from patients with suspected coronavirus symptoms.

Furthermore, the clinical circumstances are not present in these samples. Last but not least, the normal class consists of those who have neither COVID-19 nor

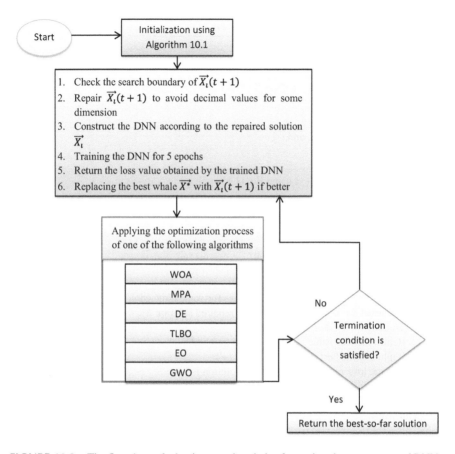

FIGURE 10.2   The flowchart of adapting metaheuristics for tuning the parameters of DNN.

pneumonia as their primary diagnosis. We are not assuming that a patient who has CXR imaging that appears "normal" does not have any developing diseases. Finally, Figure 10.3 depicts the count of the images for each class, and Figure 10.4 depicts the normal, pneumonia, and COVID-19 X-ray images.

10.4 NORMALIZATION

Normalization can be further simplified by using the min-max method. During this normalization, the minimum value of the data is changed to zero, while the greatest value is changed to one [7]. All of the values are contained inside the interval [0, 1]. After the pixel values have been subjected to this normalization, applying methods based on thresholds is simplified as a result of this. The mathematical model of this technique is as follows:

$$MinMax = \frac{x - min}{max - min}$$ (10.4)

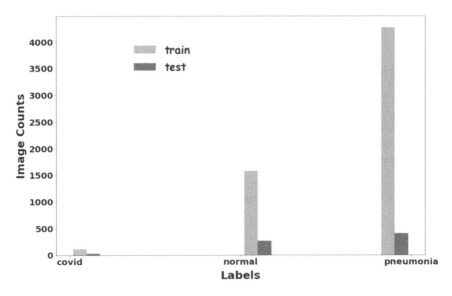

FIGURE 10.3 Image count for classes (covid, normal, and pneumonia).

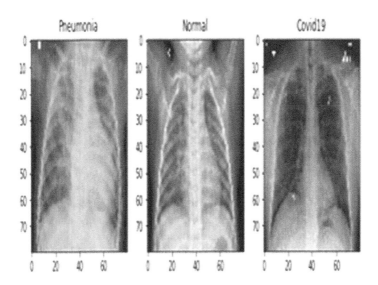

FIGURE 10.4 Images for classes (covid, normal, and pneumonia).

where x is the value of the pixel that we want to transform, while min and max are the minimum and maximum intensity values of the pixel, respectively. To normalize the pixel values, this formula is applied to each one. Because this method adds skewness to the data, it allows all of the values to lie within a relatively close range. This is one of the method's drawbacks, as it means that it is unable to manage outliers. However, this occurrence is quite unusual when it comes to picture datasets [7], therefore, it is employed in this chapter to normalize the X-ray images.

10.5   PERFORMANCE METRICS

The performance metrics employed in this chapter to evaluate the performance of the designed DNN model under seven investigated metaheuristic algorithms are the same as discussed in Chapter 3. Within this section, in a nutshell, these metrics, namely precision, recall, accuracy, specificity, and F1-score, will be overviewed. The mathematical model of each one of these metrics is as follows:

$$\text{Precision} = \frac{TP}{TP + FP} \tag{10.5}$$

$$\text{Recall} = \frac{TP}{TP + FN} \tag{10.6}$$

$$\text{Accuracy} = \frac{TP + TN}{TP + FP + TN + FN} \tag{10.7}$$

$$\text{F1} - \text{score} = \frac{Precision \cdot Recall}{Precision + Recal} \tag{10.8}$$

TN, TP, FN, and FP indicate the true negative, true positive, false negative, and false-positive samples for each class (normal, pneumonia, covid).

10.6   EXPERIMENTAL SETTINGS

Python as a programming language, with the Keras package and a TensorFlow backend, was utilized for training the suggested deep neural network. Seven well-established metaheuristic algorithms are used to tune the DNN while classifying Covid19 disease. Keras package enables the researchers to simply implement the neural network because it includes a neural network library constructed on top of Theano or TensorFlow [18]. Keras package supplies most of the building elements necessary to construct relatively advanced deep learning models. The basic computing infrastructure that has been utilized for the execution of DNNs and metaheuristic algorithms consists of a machine that has been configured as follows: with 32 gigabytes of random access memory (RAM), a 2.40 GHz Intel Core i7–4700MQ CPU, and Windows 10 operating system. All of the metaheuristic algorithms investigated in this chapter to estimate the hyper parameters of the deep neural network are implemented using Python, with a population size of 5, and a maximum number of iterations of 7. In addition to this, statistical analyses are utilized to validate the obtained findings. The training procedure is carried out for each updated solution of the metaheuristic algorithms that have been explored for a total of five epochs to evaluate the usefulness of the hyper parameters that have been produced. This process is carried out in five separate runs for each method, and the best solution for each algorithm that achieved a lower loss value is retrieved before the construction of the whole DNN model can begin. This model is further trained using the training dataset for 40 epochs to prevent overfitting for all previously trained models with a batch size of 32. As this problem consists of three classes, "categorical cross-entropy" is used as the loss function.

The controlling parameters of the investigated metaheuristic algorithms are set as found in the standard algorithm to ensure a fair comparison among them. Those parameters are described as follows. MPA has two main parameters: P and FADs that are set to 0.5 and 0.2, respectively. The EO algorithm has four main effective parameters: a_1, a_2, V, and GP which are set to 2, 1, 1, and 0.5, respectively. Regarding the parameters of DE, which are F and Cr, their values are of 0.5. Last but not least, WOA has a parameter, namely b, which is responsible for defining the logarithmic spiral shape, the value of this parameter within the conducted experiments in this section is of 1 according to the standard algorithm. Finally, SMA has one effective parameter: z, which is set to 0.03.

10.7 EXPERIMENTAL FINDINGS

The classification performance of the metaheuristic-based DNN models, namely GWO_DNN, DE_DNN, WOA_DNN, MPA_DNN, SMA_DNN, TLBO_DNN, and EO_DNN is displayed here for your perusal. To analyse the findings, the following criteria were used for each class (pneumonia, covid, and normal): precision, F1-score, recall, and the overall accuracy of the model. These metrics are displayed in Table 10.6. According to the findings reported in this table, the MPA_DNN obtains a classification accuracy of 98%, which is the highest score, while the DE_DNN and TLBO_DNN could reach a classification accuracy of 94% and 96% as the second and third best. The other models' accuracy ranges between 80% and 91%, which is considered moderate.

From this table, it has also been noticed that both MPA_DNN and GWO_DNN models have a recall of 83% for the COVID-19 class. This is extremely important because the model needs to identify all COVID-19 positive cases to stop the virus from spreading throughout the population. Meanwhile, using the WOA_DNN model would result in an accuracy rate of 80% when identifying confirmed positive COVID-19 patients. For more accurately predicting the confirmed positive COVID-19 patients, GWO_DNN and MPA_DNN models are preferred as they could reach a sensitivity of 90%. Furthermore, the DE_DNN, WOA_DNN, MPA_DNN, TLBO_DNN, GWO_DNN, and EO_DNN models each display a high precision value that is 100% for the COVID class accordingly. This implies that for these models, no classes were mistakenly classified as COVID from other classes, which is confirmed in the confusion matrix displayed in Figure 10.8. In contrast, for SMA_DNN, one normal case was incorrectly labelled as COVID, as shown in Figure 10.8.

In terms of the F1 score, a similar pattern can be observed. MPA_DNN's capacity to attain high sensitivity, F1-score, and precision in the normal class is one of the extremely positive outcomes. This guarantees that false-positive cases are minimized not only in the covid class, but also in the pneumonia class and may help reduce the strain on the healthcare system. The previous analysis concluded that MPA_DNN

TABLE 10.6
Classification Performance Obtained from Different DNN Under Investigated Metaheuristics

Algorithm	Labels	Precision	Recall	F1-score	Overall Accuracy	Avg Loss
GWO_DNN	Covid	1.00	0.83	0.91	**0.8829**	0.62327
	Normal	0.99	0.61	0.75		
	Pneumonia	0.79	1.00	0.88		
	Macro Avg	0.93	0.81	0.85		
DE_DNN	Covid	1.00	0.63	0.78	**0.9479**	0.53350
	Normal	0.91	0.98	0.95		
	Pneumonia	0.97	0.95	0.96		
	Macro Avg	**0.96**	0.86	0.89		
WOA_DNN	Covid	1.00	0.80	0.89	**0.9176**	0.66361
	Normal	1.00	0.81	0.89		
	pneumonia	0.87	1.00	0.93		
	Macro Avg	0.96	0.87	0.91		
MPA_DNN	Covid	1.00	0.83	0.91	**0.9826**	0.51080
	Normal	0.98	0.98	0.98		
	pneumonia	0.98	0.99	0.99		
	Macro Avg	0.99	0.94	0.96		
TLBO_DNN	Covid	1.00	0.77	0.87	**0.9667**	0.44044
	Normal	0.98	0.95	0.96		
	pneumonia	0.96	0.99	0.98		
	Macro Avg	0.98	0.90	0.94		
EO_DNN	Covid	1.00	0.40	0.57	**0.8497**	0.57680
	Normal	0.97	0.68	0.80		
	pneumonia	0.80	0.99	0.89		
	Macro Avg	0.92	0.69	0.75		
SMA_DNN	Covid	0.95	0.67	0.78	**0.7919**	0.56748
	Normal	0.96	0.50	0.66		
	pneumonia	0.74	0.99	0.85		
	Macro Avg	0.88	0.72	0.76		

could reach better overall accuracy and better macro-average (macro-avg) for recall, F1-score, and precision.

In addition to this, we visualized the loss and the accuracy of the same models while they were in the training phase. Figure 10.5 shows the training/validation loss, and Figure 10.6 shows the training/validation accuracy of GWO_DNN, DE_DNN, WOA_DNN, MPA_DNN, TLBO_DNN, EO_DNN, and SMA_DNN models, respectively. The MPA_DNN model exhibits a training process that is as smooth as possible in most cases, as shown in Figures 10.5 and 10.6. During this training process, by MPA_DNN, the loss value steadily lowers as the accuracy gradually increases. In most cases, the accuracy of training and validation over the MPA_DNN model does

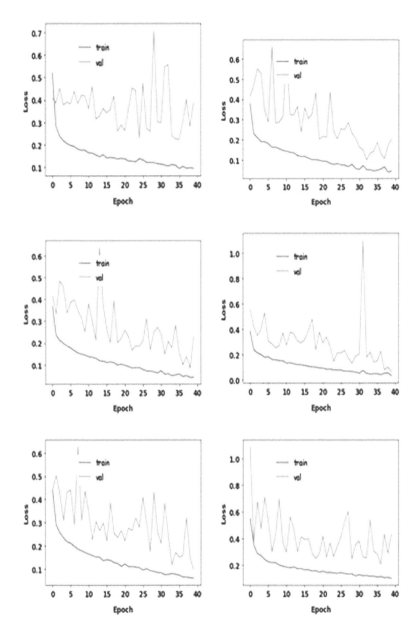

FIGURE 10.5 Loss values of DNN tuned using seven metaheuristic algorithms over the training process.

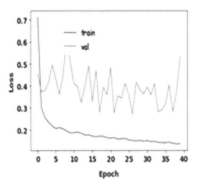

FIGURE 10.5   (Continued)

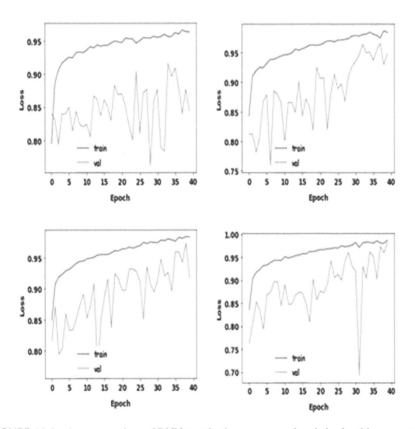

FIGURE 10.6   Accuracy values of DNN tuned using seven metaheuristic algorithms over the training process.

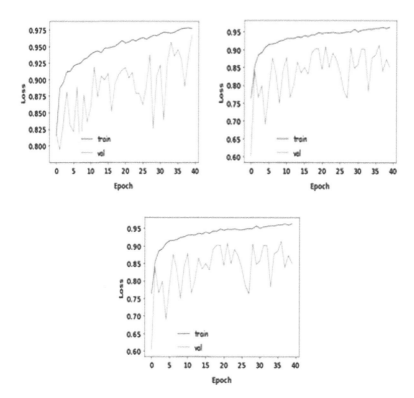

FIGURE 10.6   (Continued)

not differ significantly from one another, a characteristic that can also be observed for training and validation loss, showing that this model is not overfitting.

Finally, Figure 10.7 shows the confusion matrix of various DNN models over the test dataset. From this figure, over the test dataset, it is obvious that the models which could accurately predict the COVID-19 patients are GWO_DNN, and MPA_DNN with True positive (TP) up to 25 out of 30, and the other cases were classified as pneumonia and normal. Following GWO_DNN and MPA_DNN come WOA_DNN which could correctly predict 24 cases infected with COVID-19. Regarding the normal cases, which are incorrectly predicted as COVID-19, over the test dataset, all models except SMA_DNN do not classify any normal person as infected with COVID-19. Consequently, MPA_DNN is preferred for diagnosing COVID-19 disease from chest X-ray images to help in reducing the strain on the healthcare system.

10.8   CHAPTER SUMMARY

This chapter aims to evaluate the performance of seven different metaheuristic methods for optimizing the hyperparameters and structure design of DNN to detect

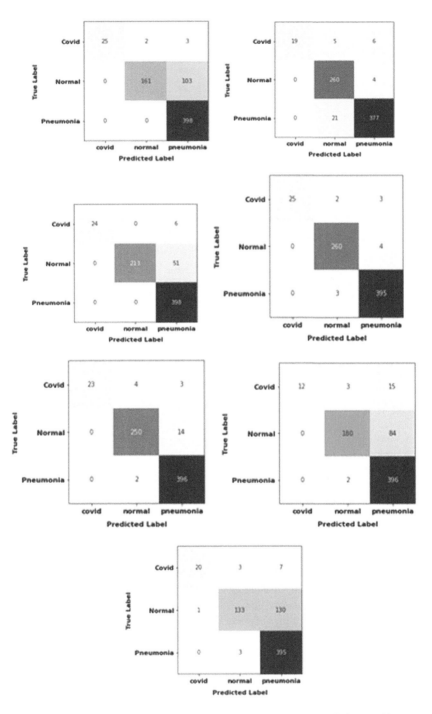

FIGURE 10.7   Confusion matrix of DNN tuned using seven metaheuristic algorithms over test dataset.

COVID-19 disease from chest X-ray images. To be more specific, a group of seven metaheuristic algorithms called GWO, DE, WOA, MPA, SMA, TLBO, and EO are utilized to estimate the number of layers, the number of neurons that make up each layer, the most effective learning, and momentum rate, as well as the most effective optimizer and activation function of DNN to improve its performance when it comes to detecting COVID-19 disease. The classification performance of the metaheuristic-based DNN models, namely GWO_DNN, DE_ DNN, WOA_DNN, MPA_DNN, SMA_DNN, TLBO_DNN, and EO_DNN, is described in this chapter for your perusal in terms of the following criteria for each class (pneumonia, covid, and normal) precision, F1-Score, recall, and the overall accuracy of the model. More specifically, MPA_DNN satisfies an overall accuracy requirement of 98%. The high values associated with sensitivity, F1-score, and precision of covid class show that this model can effectively detect positive and/or negative COVID-19 cases, hence minimizing the virus's potential to propagate across the population to as great an extent as feasible. Furthermore, one of the very encouraging results is that the MPA_DNN model can achieve high sensitivity and precision in the normal class. This ensures the minimization of false positives regarding infection classes, which has the potential to help relieve some of the strain that is placed on the healthcare system.

10.9 EXERCISES

a. Explain a few contributions of DL models for detecting COVID-related syndrome.
b. How can metaheuristic algorithms augment DL models? Provide an example.
c. How can we adapt the metaheuristic algorithm for tuning hyperparameters of DNN? Provide an example.
d. Explain the role of DL models in the healthcare system.
e. Why is a proper tuning of algorithmic parameters important? How can we do that?

REFERENCES

1. Babukarthik, R., et al., Prediction of COVID-19 using genetic deep learning convolutional neural network (GDCNN). *IEEE Access,* 2020. **8**: pp. 177647–177666.
2. Ghaderzadeh, M. and F. Asadi, Deep learning in the detection and diagnosis of COVID-19 using radiology modalities: a systematic review. *Journal of Healthcare Engineering,* 2021. 2021, Article ID: 6677314.
3. Alzubaidi, M., et al., Role of deep learning in early detection of COVID-19: Scoping review. *Computer Methods and Programs in Biomedicine Update,* 2021. **1**: p. 100025.
4. Al-Waisy, A., et al., COVID-DeepNet: hybrid multimodal deep learning system for improving COVID-19 pneumonia detection in chest X-ray images. *Computers, Materials and Continua,* 2021. **67**(2): pp. 2409–2429.
5. WHO, WHO coronavirus disease (COVID-19) dashboard. 2020. WHO.
6. Kaur, M., et al., Metaheuristic-based deep COVID-19 screening model from chest X-ray images. *Journal of Healthcare Engineering,* 2021. **2021**. Article ID: 8829829.
7. Riaz, M., et al., Metaheuristics based COVID-19 detection using medical images: A review. *Computers in Biology and Medicine,* 2022: p. 105344.

8. El-Kenawy, E.-S.M., et al., Advanced meta-heuristics, convolutional neural networks, and feature selectors for efficient Covid-19 x-ray chest image classification. *IEEE Access,* 2021. **9**: pp. 36019–36037.

9. Nematzadeh, S., et al., Tuning hyperparameters of machine learning algorithms and deep neural networks using metaheuristics: A bioinformatics study on biomedical and biological cases. *Computational Biology and Chemistry,* 2022. **97**: p. 107619.

10. Tuba, E., et al., Convolutional Neural Networks Hyperparameters Tuning, in *Artificial Intelligence: Theory and Applications*. 2021, Springer. pp. 65–84.

11. Bibaeva, V. Using metaheuristics for hyper-parameter optimization of convolutional neural networks. In *2018 IEEE 28th International Workshop on Machine Learning for Signal Processing (MLSP)*. 2018. IEEE.

12. Gaspar, A., et al., Hyperparameter Optimization in a Convolutional Neural Network Using Metaheuristic Algorithms, in *Metaheuristics in Machine Learning: Theory and Applications*. 2021, Springer. pp. 37–59.

13. De Anda-Suárez, J., et al., A novel metaheuristic framework based on the generalized Boltzmann distribution for COVID-19 spread characterization. *IEEE Access,* 2022.

14. Shankar, K., et al., Deep learning and evolutionary intelligence with fusion-based feature extraction for detection of COVID-19 from chest X-ray images. *Multimedia Systems,* 2021: pp. 1–13.

15. Khan, M.A., et al., COVID-19 case recognition from chest CT images by deep learning, entropy-controlled firefly optimization, and parallel feature fusion. *Sensors,* 2021. **21**(21): p. 7286.

16. Prasanth, S., et al., Forecasting spread of COVID-19 using google trends: A hybrid GWO-deep learning approach. *Chaos, Solitons & Fractals,* 2021. **142**: p. 110336.

17. Makris, A., I. Kontopoulos, and K. Tserpes. COVID-19 detection from chest X-Ray images using Deep Learning and Convolutional Neural Networks. In *11th Hellenic Conference on Artificial Intelligence*. 2020.

18. Gulli, A. and S. Pal, *Deep Learning with Keras*. 2017, Packt Publishing Ltd.

11 Intrusion Detection System for Healthcare System Using Deep Learning and Metaheuristics

11.1 INTRUSION DETECTION SYSTEM

Because more and more applications and domains need to make use of the internet, recent times have seen an increase in both the number of packets that need to be moved and the quantity of work that has to be done on the network. As a result, even though there are various mechanisms for network security, such as the firewall system, which is an effective protection and mitigation system, the most precious information still poses a risk. Once the information has been transferred, the firewall systems will prevent unauthorized access into the networks but will be unable to follow the surveillance. It would not be able to identify any kind of danger that was trying to get through it. The network needs to be safeguarded by an intrusion detection system (IDS) after that so that it may be managed effectively [1].

The recent growth in communications and information technologies, such as the Internet of Things (IoT), has significantly outpaced the conventional methods of detecting the environments that are close by. IoT has made it easier to create devices and infrastructures that can enhance people's lives [2]. IoT is one of the computer industry's technologies that is expanding at the quickest rate. IoT devices can be transformed into smart objects by leveraging its key technologies, such as pervasive and ubiquitous computing, embedded devices, communication technologies, sensor networks, Internet protocols, and AI-based applications [3, 4].

Recent revolutionary advancements in building IoT systems have enabled the developed healthcare monitoring systems with low-power and low-cost sensors. In recent years, these sensors have seen widespread application to permit remote patient monitoring. As a result, doctors' requirement to be physically present in the field has been eliminated [5]. A wide variety of medical applications, including early diagnosis, real-time monitoring, and medical crises, can now be supported by recent developments in IoT and wireless communications in an efficient manner [5]. Intrusion Systems are essential in e-healthcare systems because patient medical records must be kept secure, private, and accurate. Any alteration to the real patient data may result in severe issues, such as a false diagnosis or a delay in contact [5, 6].

DOI: 10.1201/9781003325574-11

In the past decade, machine learning (ML) and deep learning (DL) approaches have been implemented in cybersecurity applications for hybrid network analysis, which encompasses both misuse detection and anomaly detection. Misuse detection is used to identify known assaults based on their signatures, whereas anomaly detection is used to recognize any aberrant network behaviour. Using ML and DL to manage security risks in healthcare systems are the most prom Algorithmng methods for combating previously unidentified threats [7, 8].

In order to safeguard IoT systems from cyber threats, they need an additional layer of protection, which could be achieved using intrusion detection systems (IDSs). Several studies have attempted to characterize IDSs based on ML and DL techniques for safeguarding against IoT networks or corrupted IoT devices. These studies examined IDS research for wireless sensor networks [3], cloud-based IoT systems [3], cyber-physical systems [3], and mobile ad hoc networks (MANETs) [3]. Classical IDS approaches are less efficient or insufficient for the security of IoT systems due to their distinctive properties, such as limited energy, pervasiveness, heterogeneity, restricted bandwidth capacity, and global connectivity. Techniques based on machine learning (ML) and deep learning (DL) have lately earned credibility in a successful application for the detection of network intrusions, including IoT networks. ML/DL-based algorithms can capture both normal and abnormal activity in IoT environments. Capturing and analysing IoT devices and network information can reveal normal trends. Any deviation from these acquired normal patterns can be utilized to identify aberrant behaviour. In addition, ML/DL-based approaches have been evaluated for their ability to forecast future or zero-day threats. Therefore, ML/DL-based algorithms offer solid security protocols for the design of IoT devices and network security [3].

Thamilarasu et al. [9] suggested a wireless body area network-based intrusion detection system for the internet of medical things (WBAN). They created a hospital network topology and ran comprehensive experiments with various subsets of the Internet of Medical Things, such as wireless body area networks and other linked medical devices. For network and device-level intrusion detection, the system achieves 99.6 and 98.2% accuracy, respectively. Subas et al. [10] proposed a machine learning-based intrusion detection system with complex attack data. They concluded that the bagging ensemble technique and random forest (RF) outperform other classifier models with 97.67% accuracy.

Iwendi et al. [11] proposed an intrusion detection system in the Security of Things (SoT) paradigm for smart healthcare, which will continue to have an impact on medical infrastructures. They employed the NSL-KDD dataset for validating the performance of various machine learning models: naive Bayes (NB), and logistic regression classifiers. To maximize the functionality of their approach, they combined a weighted genetic algorithm with RF to get a high detection rate while ignoring false positives and true negatives. The combination of the genetic algorithm and RF models could achieve an accuracy rating of 98.81%. Benedetto Serinellia et al. [12] validate open-source intrusion detection datasets for the detection of well-known zero-day attacks. They developed their predictors using the KDD99, NSL-KDD, and CIC-IDS-2018 datasets, providing a comparative case study among them in their research. On the NSL-KDD datasets, they achieved 99.97% accuracy.

As discussed in the previous chapters, the DL and ML techniques are trained using the gradient-based optimizers to find the near-optimal weights which could accurately classify the training and testing datasets. But unfortunately, these optimizers are prone to falling into local minima, especially for the non-convex optimization problems which have multiple local optima (or other stationary points) that are not necessarily global optima. Therefore, in this chapter, we will investigate the performance of seven stochastic algorithms, also known as metaheuristic algorithms, which have strong properties to avoiding stuck into local minima, to train two well-established deep learning models, namely DNN, and LSTM for detecting the network intrusions. The experiments conducted in this chapter are implemented using DNN and LSTM in the Jupyter notebook using Python 3.10, Tensorflow, Keras, and Scikit learn [13, 14].

In general, the following is how this chapter is structured: The feature scaling methods will be presented at the beginning of this chapter. Then, the one-hot-encoding strategy is described. Afterwards, the implementation of the deep learning model trained using the Keras library's metaheuristic algorithms will be discussed. Last but not least, experimental results based on various performance indicators are presented to demonstrate the efficacy of seven investigated metaheuristics. Finally, a summary of the chapter is provided.

11.2   FEATURE SCALING

A procedure that takes place during the preprocessing of data is termed feature scaling. This stage is also known as standardization. The process is carried out to normalize the data within a specific range. The acceleration of the algorithm's calculations can be helped by scaling the features. Experiments conducted with Keras, Scikit learn, and deep learning all have one condition in common: the data's features must be scaled. This experiment made use of a dataset that contained variables that were measured using a variety of scales. As a result, feature scaling is applied to the dataset to transform the feature vector into a format better suited for deep learning strategies. There are numerous available scalers for scaling the dataset's features, the most commonly used ones are StandardScaler(), MinMaxScaler(), and RobustScaler() [15].

StandardScaler() [16], which is used within the experiments conducted in this chapter modifies the dataset so that the mean value of the resulting distribution is zero and the standard deviation is one. Subtracting the mean value from the original value and dividing by the standard deviation yields the transformed value. The following formula is used to transform the features using the StandardScaler():

$$z = \frac{x - \mu}{\sigma} \tag{11.1}$$

where z is the normalized value of the feature, x is the original value of the same feature, μ indicates the mean and σ stands for the standard deviation.

MinMaxScaler() [17] adjusts the data so that every value inside the dataset is ranged between 0 and 1. The mathematical equation used by the MinMaxScaler () to normalize the dataset is formulated as follows:

$$z = \frac{x - min}{max - min}$$ (11.2)

RobustScaler () [16] eliminates the median and scales the data according to the 25th to 75th quartile. Compared to the preceding scalers, the transformed values of the dataset are significantly greater. Under this scaler, the dataset values are modified to fall within the interval [2, 3]. This scaler is similar to MinMaxScaler() but utilizes inter quartile range rather than MinMaxScaler(). The mathematical model of this method is formulated as follows:

$$z = \frac{x - Q_1(x)}{Q_3(x) - Q_1(x)}$$ (11.3)

where Q_1 and Q_3 are the first and third quartile.

11.3 ONE-HOT-ENCODING

One-hot-encoding is transforming categorical data into numerical data to prepare it for a particular ML or DL algorithm. The one-hot algorithm converts each categorical value into a new categorical column and assigns those columns a binary value of 1 or 0. Each number is represented using a binary vector. The index is designated with a 1 and all the values are zero. The following figure (Figure 11.1) presents an example to elaborate on how the one-hot-encoding method works. In this figure, we convert the column type, which has categorical values, into numerical values by creating a binary vector, having a number of cells equal to the number of categories in this column. Then, the cell which corresponds to each category in the column type is set to 1, the other cells are set to 0 as shown in Figure 11.1. This method is applied in the experiments conducted within this chapter to convert the categorical values

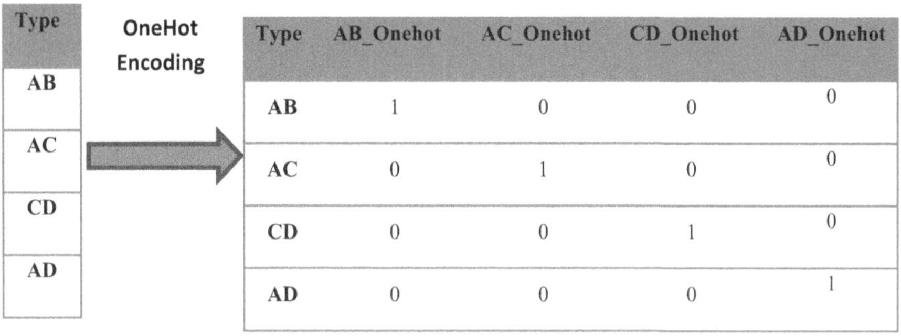

Type	OneHot Encoding	Type	AB_Onehot	AC_Onehot	CD_Onehot	AD_Onehot
AB		AB	1	0	0	0
AC		AC	0	1	0	0
CD		CD	0	0	1	0
AD		AD	0	0	0	1

FIGURE 11.1   One-hot-encoding method.

found in some column within the dataset into binary value to be suitable for both DNN and LSTM

11.4 IMPLEMENTATION

Keras is a high-level deep learning API that facilitates diverse neural networks' construction, training, evaluation, and execution. In March 2015, Keras, which Francois Chollet developed as part of a research project, was released as an open-source project. It gained popularity rapidly due to its usability, adaptability, and aesthetic appeal. It relies on a computation backend to execute the intensive calculations neural networks require. There are now three well-known open source backends: TensorFlow, Microsoft Cognitive Toolkit (CNTK), and Theano. Moreover, TensorFlow is currently bundled with its own Keras to support some helpful features, like loading and preprocessing the dataset efficiently [18]. The multilayer perceptron (MLP) model is created using the sequential library in Keras according to the following code:

Code 11.1 MLP in Keras

```
1. from Keras.layers import Dense # importing dense layer
2. from Keras.models import Sequential #importing Sequential layer
3. mlp = Sequential()
4. mlp.add(Dense(units=50, input_dim=X_train.shape[1], activation='relu'))
5. mlp.add(Dense(units=1,activation='sigmoid'))
```

Let's describe this code line by line:

- The first two lines import both dense and sequential layers.
- The third line of code creates the sequential model. This is the simplest type of Keras library for neural networks that consist just of a single stack of linked layers.
- Next, a dense hidden layer with 50 neurons, ReLU activation function, and input dimensions (input_dim) equal to the number of features in the dataset are added. Each dense layer is responsible for managing its weight matrix, containing all the connection weights between the neurons and their inputs. In addition, a vector of bias terms is managed by this layer.
- Finally, a dense layer to present the output layer with an activation function is added to the sequential model.

After creating the sequential model, the compile() method has to be called to define the loss function, optimizer, and some metrics to be estimated during training and evaluation. The last parameter is optimal. Generally, calling the compile() method within this chapter is as follows:

Code 11.2 Calling compile() in Keras

```
1. mlp.compile(loss='binary_crossentropy', optimizer='adam', metrics=['accuracy'])
```

In this code, the 'binary_crossentropy' as a loss function is used because the classification problem tackled in this chapter is binary. Regarding the optimizer, the adam optimizer described in the previous chapter will be set to train the model. Finally, the accuracy metric during training and evaluation is measured to determine the performance of the model. After describing how to code the MLP model in Keras, it's the turn to show how to implement the deep neural network, known as LSTM, in Keras. LSTM model is created using the same steps as the MLP model by replacing the dense layers within the hidden layers with the LSTM layer as shown in Code 11.3:

Code 11.3 LSTM in keras

```
1. from keras.layers import Dense # importing LSTM layer
2. from keras.models import Sequential #importing Sequential layer
3. mlp = Sequential()
4. mlp.add(LSTM(50, X_train.shape[1]))
5. mlp.add(Dense(units=1,activation='sigmoid'))
```

After creating both LSTM and MLP models, let's describe how they could be trained using the metaheuristic algorithms under the Keras library. In the beginning, all the investigated metaheuristic algorithms: WOA, SMA, MPA, DE, TLBO, GWO, and EO will be coded in python like WOA coded in Code 11.4 as a guide to be followed for the other algorithms. **Code 11.4** has similar codes to those described in Chapter 1, but some of the other lines need to be described in detail. Starting with Line 2, this line calls an initialization method to create a number, N, of solutions, referred to as population, where each solution consists of the dimensions of an optimization problem. For example, if we need to find the smallest value of a mathematical function having six variables, then a solution with six dimensions will be created. The optimization problem tackled in this chapter has a number of dimensions which is dynamically changed according to the number of layers in both DNN and LSTM, in addition to the number of features in the datasets. **Code 11.5** has been employed to compute the number of dimensions for each model without looking into the number of layers and neurons given by the researchers. In this code, we get all the weights within the model, and then count these weights to determine the number of dimensions that need to be optimized using the stochastic algorithms. After determining the number of dimensions, the initialization method (Code 11.6) is called to create the initial population, and evaluate each individual in this population. Note that the Fit variable in Code 11.6 represents an array of a global vector, including the fitness value of each individual

Afterward, a variable, bF, is created to contain the best-so-far fitness value obtained by WOA and a vector, bPos, to include the best-so-far positions that fulfil this best fitness value. Lines 8–10 in Code 11.4 update both bF and bPos according to the fitness value of each individual compared to bF. Line 11 starts the optimization process of WOA to update the initial population to reach better outcomes. Finally, the objective function used to compute the fitness value of each individual is listed in Code 11.7. This code takes a vector including the solution of each individual as a parameter. Then, this solution is divided into a number of parts equal to the number of layers in the model, such that each part contains the weights of the corresponding layer. These parts are stored in the array, namely ws. Last but not least, this array is set to the model using the function set_weights(ws) built in the sequential model. Finally, the evaluation function is coded into the sequential model and takes two main parameters: X_train and y_train; X_train contains the training data which is used to train the model, while y_train includes the true label of each record in X_train.

Code 11.4 WOA in python

```
1.   def WOA():
2.     dim=GetDim()
3.     initialization(dim, N)
4.     t=0
5.     bF=float('inf')
6.     bPos=[];
7.     for i in range(N):
8.     if Fit[i]<bF:
9.        bF=Fit[i]
10.       bPos=copy.copy(X[i,:]);
11.  while t<Max_iter:
12.    a=2-t*((2.0)/Max_iter);
13.
14.    # a2 linearly dicreases from -1 to -2 to calculate t in Eq. (3.12)
15.    a2=-1+t*((-1)/Max_iter);
16.
17.    # Update the Position of search agents
18.    for i in range(N):
19.    r1=np.random.uniform(); # r1 is a random number in [0,1]
20.    r2=np.random.uniform(); # r2 is a random number in [0,1]
21.    A=2*a*r1-a;
22.    C=2*r2;
23.    b=1;
24.    l=(a2-1)*np.random.uniform()+1;
25.    p = np.random.uniform();
26.    for j in range(dim):
27.      if p<0.5:
28.        if abs(A)>=1:
29.            rand_leader_index = int(N*np.random.uniform());
30.            X_rand = X[rand_leader_index,:];
31.            D_X_rand=abs(C*X_rand[j]-X[i,j]); # Equation (2.7)
32.            X[i,j]=X_rand[j]-A*D_X_rand; # Equation (2.8)
33.        elif abs(A)<1:
34.            D_Leader=abs(C*bPos[j]-X[i,j]); # Equation (2.1)
35.            X[i,j]=bPos[j]-A*D_Leader; # Equation (2.2)
36.      elif p>=0.5:
37.            distance2Leader=abs(bPos[j]-X[i,j]);
38.            # Equation (2.5)
39.            X[i,j]=distance2Leader*math.exp(b*l)*math.cos(l*2*math.pi)+bPos[j];
40.    Fit[i], acc= _objective_function__(X[i,:])
41.    if Fit[i]<bF:
42.      bF=Fit[i]
43.      bPos=copy.copy(X[i,:]);
```

```
44.        bestTestResult=acc;
45.    t=t+1;
46. return bestTestResult;
```

Code 11.5 GetDim()

```
1.  def _setting_paras__():
2.     dim = 0
3.     for wei in mlp.get_weights():
4.         dim += len(wei.reshape(-1))
5.     return dim
```

Code 11.6 Initialization(dim, N)

```
1.  def initialization():
2.     for i in range(N):
3.         for j in range(dim):
4.             X[i,j]=lb[j]+np.random.random(1)*(ub[j]-lb[j])
5.         Fit[i]= _objective_function__(X[i])
```

Code 11.7 _objective_function__(solution=None)

```
1.  def _objective_function__(solution=None):
2.         weights = [rand(*w.shape) for w in mlp.get_weights()]
3.         ws = []
4.         cur_point = 0
5.         for wei in weights:
6.             ws.append(reshape(solution[cur_point:cur_point + len(wei.reshape(-1))], wei.
               shape))
7.             cur_point += len(wei.reshape(-1))
8.         mlp.set_weights(ws)
9.         test_results=mlp.evaluate(X_train, y_train, verbose=0)
10.        return test_results[0]
```

11.5   DATASET DESCRIPTION

The performance of the seven-investigated metaheuristic algorithms for training both MLP and LSTM is validated using a widely-used dataset, known as NSL-KDD. The NSL-KDD dataset was proposed to address difficulties inherent to the KDD Cup 1999 dataset, which contains an excessive number of redundant records. This dataset contains five primary intrusion categories: Normal (No intrusion), denial-of-service (DoS), Probe, U2R, and R2L. These incursion kinds are briefly described below:

- **Normal**: Networks that are not infected by any attack.
- **DoS**: This is the acronym for denial-of-service. This attack submits a flood of requests to the server, which are higher than its ability.
- **Probe**: This attack inserts a program or other device at a key juncture in a network for monitoring or collecting data about network activity.
- **User to root (U2R):** An intruder tries to obtain access to a regular user account.

TABLE 11.1
Distribution of Intrusion Classes

Attack	Number of records
Normal	67343
DOS	45927
Probe	11656
U2R	52
R2L	995

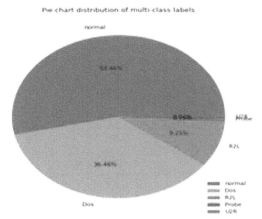

FIGURE 11.2   Pie chart distribution of multi-class labels.

- **Remote to local (R2L):** Unauthorized access coming from a different machine.

The distribution of the intrusion classes mentioned before within the NSL-KDD is described in Table 11.1. In addition, the pie chart distribution of multi-class and binary labels (normal and abnormal) is described in Figures 11.2 and 11.3, respectively. In this chapter, we will deal with this problem as a binary problem, not a multi-class problem, where the abnormal is represented with the value 0 to indicate that any type of intrusions mentioned before attacked the network, and the normal category is represented with 1 to indicate that the network is not attacked.

11.6   PERFORMANCE METRICS

Several performance metrics are utilized to evaluate the performance of the stochastic algorithms for training the MLP and LSTM models; those metrics are precision, recall, accuracy, F1-score, mean square error (MSE), and mean absolute error (MAE), and root mean square error (RMSE). The mathematical formula of those metrics is described below:

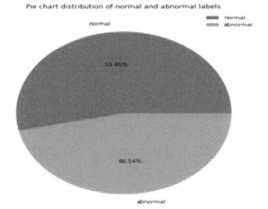

FIGURE 11.3   Pie chart distribution of normal and abnormal labels.

$$\text{Precision} = \frac{TP}{TP + FP} \qquad (11.4)$$

$$\text{Recall} = \frac{TP}{TP + FN} \qquad (11.5)$$

$$\text{Accuracy} = \frac{TP + TN}{TP + FP + TN + FN} \qquad (11.6)$$

$$\text{F1}-\text{score} = \frac{Precision \cdot Recall}{Precision + Recal} \qquad (11.7)$$

$$MSE = \frac{1}{m}\sum_{i=1}^{i=m}\left(o_i^D - d_i^D\right)^2 \qquad (11.8)$$

$$RMSE = \sqrt{\frac{1}{m}\sum_{i=1}^{i=m}\left(o_i^D - d_i^D\right)^2} \qquad (11.9)$$

$$MAE = \frac{1}{m}\sum_{i=1}^{i=m}\left|o_i^D - d_i^D\right| \qquad (11.10)$$

$$R2 = 1 - \frac{MSE}{\frac{1}{m}\sum_{i=1}^{i=m}\left(o_i^D - \bar{o}\right)^2} \qquad (11.11)$$

where TN, TP, FN, and FP indicates the true negative, true positive, false negative, and false positive samples for each class (normal, abnormal). $\bar{o}$ represents the mean of all actual values. All the performance metrics except MSE, RMSE, and MAE have to be maximized to reach better performance by the evaluated model.

Furthermore, an additional metric, known as AUC-ROC, is employed to display the performance of a classification model as an evaluation metric [19]. The receiver operating characteristic (ROC) curve illustrates the ratio of the number of true positives to the number of false positives, hence drawing attention to the sensitivity of the classifier model. The two-dimensional region that is present underneath the ROC curve is what the area under the curve (AUC) measures. The area under the curve (AUC) of a classifier is equal to the likelihood that the classifier would rank an arbitrarily selected positive example higher than it ranks an arbitrarily selected negative example. The AUC is a summary statistic that is used in conjunction with the ROC curve. AUC measures a classifier's ability to differentiate between different categories. It is assumed that the model performs better when differentiating between the positive and negative classes when AUC is higher.

11.7 EXPERIMENTAL SETTINGS

Python as a programming language, with the Keras package and a TensorFlow backend, was utilized in order to implement the MLP and LSTM models. These models are trained using seven well-established metaheuristic algorithms to investigate the metaheuristic's ability to reach the near-optimal weights which detect the network attacks. All algorithms in this chapter are implemented over a device with the following capabilities: 32 gigabytes of random access memory (RAM), a 2.40 GHz Intel Core i7–4700MQ CPU, and a Windows 10 operating system.

Most metaheuristic algorithms have two main parameters: population size and maximum iteration, which must be initialized before starting the optimization process. These parameters are set to 20 and 30, respectively, to ensure a fair comparison among algorithms investigated in this chapter. All algorithms are executed five independent times due to their stochastic nature, and the obtained outcomes within these times are presented in the next section using various performance metrics mentioned before.

To ensure a fair comparison among the researched metaheuristic algorithms, their controlling parameters are set to match those of the standard algorithm. Following is a description of these parameters: MPA has two primary parameters: P and FADs, which are both set at 0.5 and 0.2, respectively; The EO method contains four effective parameters: a1, a2, V, and GP, which are set to 2, 1, 1, and 0.5, respectively; Regarding the parameters of DE, which are F and Cr, their values are 0.5; Last but not least, WOA has a parameter, b, which is set to 1 as found in the standard algorithm; Lastly, SMA has a single effective parameter: z, which is set to 0.03.

11.8 EXPERIMENTAL FINDINGS

This section will investigate the performance of seven metaheuristic algorithms: GWO, WOA, MPA, SMA, DE, TLBO, and EO for training two well-known deep learning models: MLP and LSTM as an attempt to overcome the local optima problem which the gradient-based optimizers might cause. The rest of this section will present the results obtained by each model and make extensive comparisons among models using various performance metrics to find the most effective metaheuristic algorithm.

TABLE 11.2
Classification Performance Obtained from MLP Trained, Under-Investigated Metaheuristics

Algorithm	Labels	Precision	Recall	F1-score	Accuracy	MAE	MSE	RMSE	R2
GWO_MLP	Normal	0.92	0.95	0.93	92.8081	0.15236	0.06177	0.24854	75.4839
	Abnormal	0.94	0.91	0.92					
DE_MLP	Normal	0.92	0.95	0.93	92.9827	0.09338	0.05113	0.22612	79.4679
	Abnormal	0.94	0.91	0.93					
WOA_MLP	Normal	0.87	0.87	0.8	86.1846	0.17040	0.09691	0.31130	61.0813
	Abnormal	0.85	0.8	0.85					
MPA_MLP	Normal	0.93	0.93	0.93	92.9002	0.09446	0.05032	0.22434	79.7968
	Abnormal	0.92	0.92	0.92					
TLBO_MLP	Normal	**0.95**	0.96	**0.96**	**95.2848**	**0.07801**	**0.03630**	**0.19054**	**85.4366**
	Abnormal	0.95	**0.95**	**0.95**					
EO_MLP	Normal	0.95	0.92	0.94	93.3765	0.09106	0.05583	0.23630	77.8446
	Abnormal	0.91	**0.95**	0.93					
SMA_MLP	Normal	0.92	**0.97**	0.95	94.0756	0.14841	0.05636	0.23740	77.5015
	Abnormal	**0.96**	0.91	0.93					

11.8.1 METAHEURISTICS-BASED MLP MODELS

The classification performance of the metaheuristic-based MLP models, namely GWO_MLP, DE_MLP, WOA_MLP, MPA_MLP, SMA_MLP, TLBO_MLP, and EO_MLP is displayed here for your perusal. In order to analyse the findings, the following criteria were used for each class (normal, abnormal): precision, F1-score, and recall, and the total evaluation of all classes was done by seven other metrics: overall accuracy, MAE, MSE, RMSE, R2, convergence curve, average fitness value, and ROC. The results of these metrics are displayed in Table 10.6, which reports that TLBO_MLP is the best because it could overcome the other algorithms for all the performance metrics. According to the findings reported in this table, the TLBO_MLP obtains an overall accuracy of 95% as the best one, followed by SMA_MLP with accuracy up to 94%, while the WOA_MLP is the worst performed one. Generally, this table shows that TLBO_MLP could reach a better outcome for most performance metrics; hence, it is the most preferred one for detecting network intrusions that might attempt to attack the networks.

AUC_ROC metric was used to evaluate the efficiency of various investigated models. The results of this metric for each model are presented in Figure 11.4, which

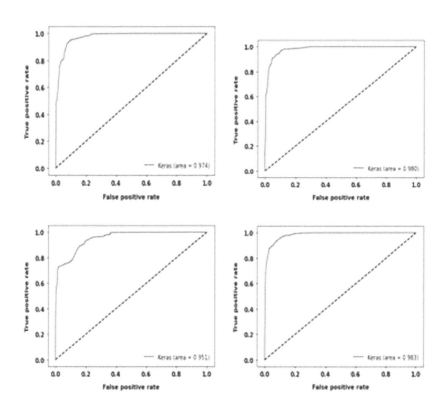

FIGURE 11.4 AUC_ROC comparison of MLP trained under various metaheuristic algorithms.

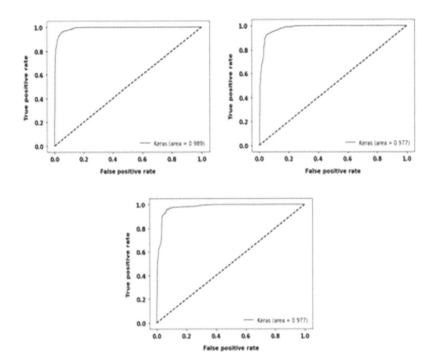

FIGURE 11.4   (Continued)

shows the efficiency of the TLBO_MLP model because it could rank as the best one in terms of AUC value (area=0.989), followed by MPA_MLP as the second best one, while WOA_MLP is the worst one. Additionally, the average convergence curve obtained by each investigated metaheuristic algorithm during training the MLP models for five independent times is computed and presented in Figure 11.5 which reveals the superiority of TLBO_MLP compared to all the other compared ones. Finally, the average fitness value obtained by each algorithm within five independent times are computed and presented in Figure 11.6. Inspecting this figure shows that TLBO_MLP could reach the lowest average fitness value of 0.142, followed by DE_ MLP and MPA_MLP as the second and third best one, respectively, while WOA_ MLP is the worst one.

11.8.2   METAHEURISTICS-BASED LSTM MODELS

The classification performance of the metaheuristic-based LSTM models, namely GWO_LSTM, DE_ LSTM, WOA_LSTM, MPA_LSTM, SMA_LSTM, TLBO_ LSTM, and EO_LSTM is displayed here for your perusal. The results of the perform-ance metrics for each model are displayed in Table 10.6, which reports that TLBO_ LSTM comes in the first rank in terms of accuracy, R2, MAE, MSE, RMSE, F1-score,

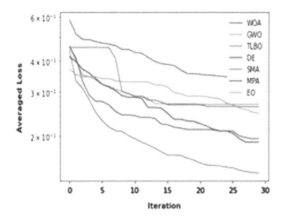

FIGURE 11.5 Convergence curve of algorithms when training MLP.

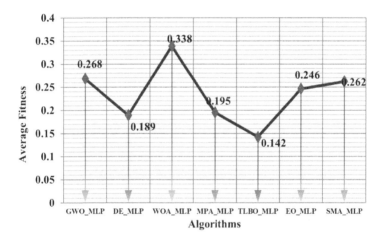

FIGURE 11.6 Average fitness value of algorithms when training MLP.

recall for the abnormal class, and precision for the normal class. Numerically comparison, TLBO_LSTM could reach an accuracy of 95.3419 as the best-obtained accuracy, followed by DE_LSTM as the second best one, while the third and fourth best models are MPA_LSTM and EO_LSTM, respectively. The worst performed one in terms of accuracy is SMA_LSTM with a value up to 86.17. TLBO_LSTM could reach better value for the metrics: MAE, MSE, RMSE, and R2 with values of 0.06599, 0.03458, 0.18596, and 86.1300, respectively.

The AUC ROC metric was applied in order to judge how effective the various models under investigation were. The results of this metric for each model are presented in Figure 11.7. TLBO_LSTM model was able to rank as the best in terms of AUC value (area=0.988), followed by DE_LSTM as the second best one, and SMA_LSTM is the model that performed the worst. Additionally, the average convergence

TABLE 11.3
Classification Performance Obtained from Different LSTM Trained Under Investigated Metaheuristics

Algorithm	Labels	Precision	Recall	F1-score	Accuracy	MAE	MSE	RMSE	R2
GWO_LSTM	Normal	0.91	0.87	0.89	88.4708	0.18966	0.08065	0.28400	68.9661
	Abnormal	0.86	0.90	0.88					
DE_LSTM	Normal	0.93	0.96	0.94	93.9702	0.12042	0.04684	0.21643	81.2047
	Abnormal	0.95	0.92	0.93					
WOA_LSTM	Normal	0.86	0.98	0.92	90.3219	0.14255	0.07095	0.26636	71.9521
	Abnormal	0.97	0.81	0.89					
MPA_LSTM	Normal	0.94	0.95	0.94	93.8845	0.07870	0.04807	0.21925	80.7016
	Abnormal	0.94	0.93	0.93					
TLBO_LSTM	Normal	**0.95**	0.97	**0.96**	**95.3419**	**0.06599**	**0.03458**	**0.18596**	**86.1300**
	Abnormal	0.96	**0.94**	**0.95**					
EO_LSTM	Normal	0.93	0.95	0.94	93.3034	0.12103	0.052144	0.22835	79.0779
	Abnormal	0.94	0.92	0.93					
SMA_LSTM	Normal	0.80	**0.98**	0.88	86.1751	0.19869	0.104012	0.32250	62.4891
	Abnormal	**0.97**	0.73	0.83					

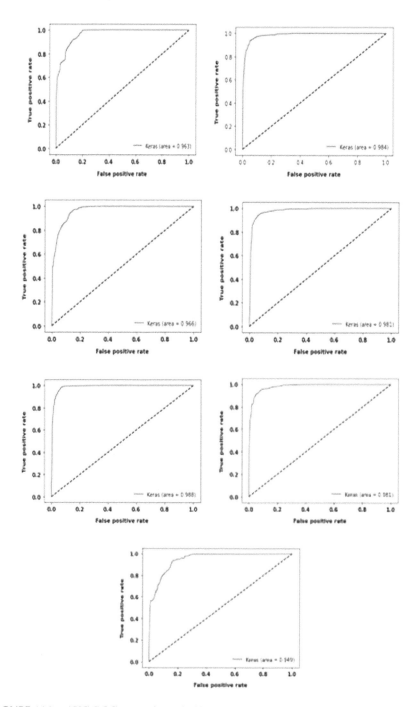

FIGURE 11.7   AUC_ROC comparison of LSTM trained under various metaheuristic algorithms.

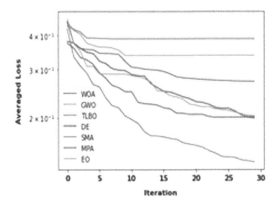

FIGURE 11.8   Convergence curve of algorithms when training LSTM.

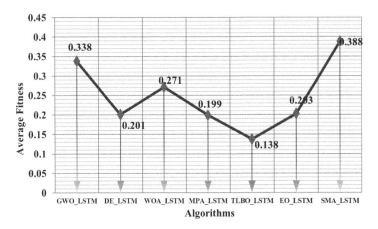

FIGURE 11.9   Average fitness value of algorithms when training LSTM.

curves produced by each analysed metaheuristic algorithm when training the LSTM models for five independent times are computed and displayed in Figure 11.8, demonstrating that TLBO_LSTM is superior to all of the other ones that were compared to it. In conclusion, the average fitness value obtained by each method over the course of five separate runs has been computed and is shown in Figure 11.9. Taking a closer look at this data reveals that TLBO_LSTM was able to achieve the lowest average fitness value of 0.138. MPA_LSTM and DE_LSTM followed TLBO_LSTM as the second and third best ones, respectively; SMA_LSTM is the worst one with an average fitness value of 0.388.

This section illustrates how effective TLBO is for training LSTM to increase its effectiveness in identifying intrusions that may pose a hazard to networks. Additionally, the prior section demonstrates that TLBO is capable of training the MLP model more effectively than any of the other investigated metaheuristic algorithms.

In addition, we have reached the conclusion that, due to the fact that the performance of the analysed algorithms varies, it is possible that another metaheuristic algorithm might achieve a higher level of accuracy than TLBO. As a result of this, in most cases, the metaheuristic algorithms could be used to train deep learning models in order to improve their performance.

11.9   CHAPTER SUMMARY

In this chapter, we investigate the performance of seven stochastic algorithms, which are also known as metaheuristic algorithms. These algorithms have strong properties to avoid getting stuck into local minima, and we use them to train two well-established deep learning models, namely MLP and LSTM, to detect network intrusions. The Jupyter notebook, Python 3.10, Tensorflow, Keras, and Scikit learn were used to carry out the experiments conducted in this chapter. The experimental findings demonstrate that TLBO is capable of training the MLP and LSTM models more effectively than any of the other investigated metaheuristic algorithms. Since the performance of the investigated algorithms varies, we have also concluded that another metaheuristic algorithm may attain a greater level of accuracy than TLBO. As a result, metaheuristic methods could be used to train deep learning models to enhance their performance in the majority of instances.

11.10   EXERCISES

a.   What is the Internet of Things (IoT). How has the advancement of IoT contributed to a better healthcare management system?

b.   How can metaheuristic algorithms augment DL models? Provide an example.

c.   What is metaheuristic algorithms' contribution to intrusion detection in the healthcare system?

d.   Why is a proper tuning of algorithmic parameters important? How can we do that?

e.   Why is intrusion detection important for the healthcare system?

REFERENCES

1. Pande, S., A. Khamparia, and D. Gupta, An intrusion detection system for healthcare system using machine and deep learning. *World Journal of Engineering,* 2021. **19**(2): pp. 166–174.

2. Ray, S., et al., The changing computing paradigm with internet of things: A tutorial introduction. *IEEE Design & Test,* 2016. **33**(2): pp. 76–96.

3. Asharf, J., et al., A review of intrusion detection systems using machine and deep learning in internet of things: Challenges, solutions and future directions. *Electronics,* 2020. **9**(7): p. 1177.

4. Khan, R., et al. Future internet: The internet of things architecture, possible applications and key challenges. In *2012 10th International Conference on Frontiers of Information Technology.* 2012. IEEE.

5. Hady, A.A., et al., Intrusion detection system for healthcare systems using medical and network data: A comparison study. *IEEE Access,* 2020. **8**: pp. 106576–106584.

6. Kumaar, M.A., et al., A hybrid framework for intrusion detection in healthcare systems using deep learning. *Frontiers in Public Health,* 2021. **9**: 824898.
7. Buczak, A.L., E. Guven, A survey of data mining and machine learning methods for cyber security intrusion detection. *IEEE Communications Surveys & Tutorials,* 2015. **18**(2): pp. 1153–1176.
8. Atzori, L., A. Iera, and G. Morabito, The internet of things: A survey. *Computer Networks,* 2010. **54**(15): pp. 2787–2805.
9. Thamilarasu, G., A. Odesile, and A.J.I.A. Hoang, An intrusion detection system for internet of medical things. *IEEE Access,* 2020. **8**: pp. 181560–181576.
10. Subasi, A., et al. Intrusion detection in smart healthcare using bagging ensemble classifier. In *International Conference on Medical and Biological Engineering.* 2021. Springer.
11. Iwendi, C., et al., Security of things intrusion detection system for smart healthcare. *Electronics,* 2021. **10**(12): p. 1375.
12. Serinelli, B.M., A. Collen, and N.A. Nijdam, On the analysis of open source datasets: validating IDS implementation for well-known and zero day attack detection. *Procedia Computer Science,* 2021. **191**: pp. 192–199.
13. Gulli, A. and S. Pal, *Deep Learning with Keras.* 2017, Packt Publishing Ltd.
14. Shukla, N. and K. Fricklas, *Machine Learning with TensorFlow.* 2018, Manning Greenwich.
15. Thara, D., B. PremaSudha, and F. Xiong, Auto-detection of epileptic seizure events using deep neural network with different feature scaling techniques. *Pattern Recognition Letters,* 2019. **128**: pp. 544–550.
16. McClarren, R.G., A. 1 Scikit-Learn. *Machine Learning for Engineers,* 2021: p. 239.
17. Powers, D.M., *Evaluation: from precision, recall and F-measure to ROC, informedness, markedness and correlation.* 2020 School of Informatics and Engineering, Flinders University.
18. Douglass, M.J., *Book Review: Hands-on Machine Learning with Scikit-learn, Keras, and Tensorflow, by Aurélien Géron.* 2020, Springer.
19. Narkhede, S.J.T.D.S., Understanding auc-roc curve. *Towards Data Science,* 2018. **26**(1): pp. 220–227.

Index